Mental Health for Hispanic Communities

Ruby Castilla-Puentes • Tatiana Falcone
Editors

Mental Health for Hispanic Communities

A Guide for Practitioners

Springer

Editors
Ruby Castilla-Puentes
Center for Clinical and Translational Science and Training
University of Cincinnati
Cincinnati, OH, USA

Tatiana Falcone
Cleveland Clinic
Lerner College of Medicine
Cleveland, OH, USA

ISBN 978-3-031-13194-3 ISBN 978-3-031-13195-0 (eBook)
https://doi.org/10.1007/978-3-031-13195-0

This Springer imprint is published by the registered company Springer Nature Switzerland AG
The registered company address is: Gewerbestrasse 11, 6330 Cham, Switzerland

Foreword

We are pleased to write this Foreword to our *Mental Health for Hispanic Communities: A Guide for Practitioners*.

The Hispanic/Latinx community in the U.S. is very diverse, including people from many different nations and regions of the world. As with any community, the mental health needs and experiences of Hispanic/Latinx people vary among subgroups [1]. While there is great diversity within this community, there are some shared cultural factors that connect people regardless of ancestry or national origin [2]. A large portion of this community speaks the Spanish language. There is also a shared connection of religious affiliations, strong family bonds, connections to extended networks, and a resilient approach to life and work. When mental health is not commonly or openly talked about, people seeking treatment may have limited knowledge and comfort with different types of therapy and psychiatric medications [3].

Clinicians should use a Culturally and Linguistically Competent Care to engage Hispanic/Latinx patients in treatment planning. Incorporating education, symptom monitoring, and engagement with community resources can be important to support a person's decision to start therapy or psychiatric medication [4].

In this book, we present an accessible approach to diagnosing and treating Hispanic/Latinx psychiatric patients. We begin by presenting an overview of the history and address the key factors influencing mental health in Hispanic communities: The role of religion, culture, and stigma in mental health.

Perhaps the most important reason of all for developing psychiatric expertise in Hispanic/Latinx population is that the lack of a culturally oriented assessment makes the management of psychiatric illness—that is the diagnosis and treatment—much more difficult.

We have produced a practical and readable guide to the diagnosis and treatment of Hispanic/Latinx patients with psychiatric disorders. We encourage you to read it and take it to heart. Your Hispanic/Latinx patients and families suffering from psychiatric disorders will thank you for it.

Our book will not eliminate the universal experience of non-Hispanic clinicians feeling like playactors, but it will go a long way toward providing a practical framework that the clinician can use to anchor the early stages of learning how to work with Hispanic/Latinx patients.

References

1. Office of the Surgeon General (US); Center for Mental Health Services (US); National Institute of Mental Health (US). Mental Health: Culture, Race, and Ethnicity: A Supplement to Mental Health: A Report of the Surgeon General. Rockville (MD): Substance Abuse and Mental Health Services Administration (US); 2001 Aug. Chapter 6 Mental Health Care for Hispanic Americans. Available from: https://www.ncbi.nlm.nih.gov/books/NBK44247/
2. Center for Substance Abuse Treatment (US). Improving Cultural Competence. Rockville (MD): Substance Abuse and Mental Health Services Administration (US); 2014. (Treatment Improvement Protocol (TIP) Series, No. 59.) 1, Introduction to Cultural Competence. Available from: https://www.ncbi.nlm.nih.gov/books/NBK248431/
3. Dixon LB, Holoshitz Y, Nossel I. Treatment engagement of individuals experiencing mental illness: review and update [published correction appears in World Psychiatry. 2016 Jun;15(2):189]. World Psychiatry. 2016;15(1):13–20. https://doi.org/10.1002/wps.20306
4. Valdez CR, Dvorscek MJ, Budge SL, Esmond S. Provider Perspectives about Latino Patients: Determinants of Care and Implications for Treatment. Couns Psychol. 2011;39(4):497–526.

Cincinnati, OH, USA — Ruby Castilla-Puentes
Cleveland, OH, USA — Tatiana Falcone

Preface

We wrote this book for practitioners. For psychiatrists, psychologists, physicians, nurse practitioners, physician assistants, and social workers—and for students, particularly for their family medicine, internal medicine, pediatrics, and psychiatry clerkships. Our aim is to provide help for clinicians in providing mental health care to Hispanic/Latinx patients and to guide educators in their training of students, residents, and practitioners in these areas.

There is an urgent need for all practitioners to become skilled in caring for Hispanic/Latinx patients with mental disorders. Why? As part of the fastest growing and largest ethnic minority group, projections indicate that approximately 130 million Hispanic/Latinx will reside in the United States (U.S.) by 2050. In addition, the Hispanic/Latinx community in the U.S. is very diverse, including people from many different nations and regions of the world. Individuals of Mexican, Puerto Rican, Cuban, Central American, and South American descent have been part of the American cultural framework for centuries.

While there is great diversity within this community, there are some shared cultural factors that connect people regardless of indigenous roots, ancestry, or national origin. A large portion of this community speaks the Spanish language. Our first recommendation is very simple, if you can make every effort to interact more with people of Hispanic roots, visit a country where they only speak Spanish and make a conscious attempt to understand Hispanic culture better. There is also a shared connection of religious affiliations, strong family bonds, connections to extended networks, and a resilient approach to life and work that only Hispanic/Latinx people could understand.

Offering culturally sensitive education to providers is critical. This failure leads not only to suffering in patients and families but also to frustration and burnout as clinicians. It's hard to give care you're not trained to provide. That's why we've written this book—to offer some help. We can be more empathic with our patients, when we can understand their culture better.

For psychiatrists and mental health providers working with Hispanic/Latinx patients, exploring cultural identity may offer important information to tailor their mental health treatment. Cultural humility is necessary to provide quality care. This refers to the ability to recognize that culture plays a large role in a person's health and well-being and may sometimes affect the provider's ability to best serve their patient's needs.

However, untrained providers for culturally appropriate mental health care can have difficulty with diagnosis and managing patients from different ethnic backgrounds. It's not just mental health care that suffers when mental health problems aren't treated. With the frequent co-occurrence of a medical disease and a mental disorder, we know that medical care suffers until the psychiatric disorder is also recognized and treated effectively.

A provider who understands a patient's culture and needs will know culturally specific information. A culturally sensitive doctor would be aware of Hispanic/Latinx interpretations and would ask for more information. Greatly compounding this problem, psychiatric disorders are the most common health problem in the U.S., more common than heart disease and cancer combined. One-half of the entire U.S. population will have a mental disorder at some point during their lives, one-fourth in any given year. In the average clinic, approximately one-half of patients will have some psychiatric diagnosis.

Specifically, this book (1) describes how to identify in Hispanic/Latinx patients common core of mental disorders for which clinicians need significant mastery—e.g., depression, anxiety disorders, psychotic disorders, and substance use problems; (2) describes how to identify in Hispanic/Latinx patients less common and more severe mental disorders that require early recognition and early referral—for example, psychotic disorders, schizophrenia, and suicide; and (3) describes the main risk factors for the presence and persistence of mental health conditions in Hispanic/Latinx population (e.g., religion, acculturation, etc.).

The authors are psychiatrists and mental health professionals, and they have been trained medical students, faculty, and residents for many years. Their clinical, research, experience with Hispanic/Latinx patients will support many practitioners improve mental care for Hispanic/Latinx population.

We do not propose to make culturally competent mental health professionals in Hispanic/Latinx culture out of readers but, rather, to prepare them to manage the common mental disorders which are likely to be encountered in practice with Hispanic/Latinx patients.

Cincinnati, OH, USA — Ruby Castilla-Puentes
Cleveland, OH, USA — Tatiana Falcone

Contents

1 **The Role of Religion in Mental Health** 1
Juan R. Jaramillo

2 **A History of Mental Health Treatment in the Hispanic World** 9
Fernando Espí Forcén

3 **Mental Health Stigma in the Latinx Population: Treatment Implications and Future Directions** 17
Stacie Craft DeFreitas

4 **Psychotic Disorders in Hispanics**. 33
Tatiana Falcone, Ruby Castilla-Puentes, and Alireza Minagar

5 **Mood Disorders in Hispanic/Latinos** 49
Mauricio Tohen and Hugo Gomez Rueda

6 **Anxiety in Hispanics** 63
Pilar Lachhwani

7 **The Role of Acculturation in the Mental Health of Hispanics** 87
Renato D. Alarcón

8 **Suicidal Thoughts and Behaviors in Hispanic and Latino Communities** 105
Danielle Torres, Sharely Fred-Torres, Elizabeth Soto, and M. Mercedes Perez-Rodriguez

9 **Substance Use in the Latino Population in the United States**. 127
Youssef Mahfoud, Michael Ignatowski, and Jessica Gonzalez-Tindle

10 **Dementia and Cognitive Disorders in Geriatric Hispanic/Latinos** . . . 137
Bernardo Ng and Nancy C. Colimon-Ardila

11 **Mental Health Issues in Latinx/Hispanic Children and Adolescents** . . . 153
Tatiana Falcone and Anjali Dagar

12 Cultural Sensitive Services for Latinos 169
Andrea Mendiola-Iparraguirre, Juan Rodriguez-Guzman, and Esperanza Diaz

13 Medical Comorbidities of Hispanics with Mental Health Problems 183
David Martinez Garza, Elliott R. Weinstein, and Daniel E. Jimenez

14 Mental Health in Hispanic/Latina/Latinx Women 205
Wilma I. Castilla-Puentes, Ruby Castilla-Puentes, and Sandra Castilla-Puentes

Index 215

The Role of Religion in Mental Health 1

Juan R. Jaramillo

Vocatus atque non vocatus Deus aderit (Bidden or not bidden, God is present). Delphi Oracle. Sentence carved above Carl Jung's door at his house in Kusnacht, Switzerland.

No topic should be forbidden when it comes to mental health. Religion and spirituality (R/S) are no exceptions. According to Gallup, 64% to 87% of Americans believe in God [1]. Asking about R/S should be part of every psychiatric assessment. Just like practitioners ask about depression and panic, stress and hobbies, sexuality and intimacy, financial and work issues, among others, they should ask about R/S issues.

R/S and Psychiatry have had a difficult relation. Freud boasted about being "an infidel Jew" and called religion a collective delusion. Much has changed since then. As a Psychiatry Fellow, 25 years ago, I was taught at Memorial Sloan-Kettering Cancer Center to respect my patients' spiritual beliefs, to inquire about them and to be comfortable working with Chaplains.

Talking about the role of Religion, the Massachusetts General Hospital's book of Consult/Liaison Psychiatry says that "ignoring religious content risks omitting an important element of the psychotherapeutic armamentarium" [2]. This may reflect the perhaps more relevant role of R/S in Consult/Liaison Psychiatry, where patients face, frequently, big existential issues. However, the World Psychiatric Association, in its official statement on R/S says that "spirituality and religion are concerned with the core beliefs, values and experiences of human beings [3]. A consideration of their relevance to the origins, understanding and treatment of psychiatric disorders and the patient's attitude toward illness should therefore be central to clinical and academic psychiatry." In the same statement a number of recommendations follow, including things like "a tactful consideration of patients' religious beliefs and practices as well as their spirituality should routinely be considered and will sometimes

J. R. Jaramillo (✉)
The Sidney & Lois Eskenazi Hospital, Indianapolis, IN, USA
e-mail: juan.jaramillo@eskenazi.edu

R. Castilla-Puentes, T. Falcone (eds.), *Mental Health for Hispanic Communities*,
https://doi.org/10.1007/978-3-031-13195-0_1

be an essential component of psychiatric history taking." The document also talks about the need to include R/S competencies in the training of Residents.

An important question on this topic is how Religion and Spirituality are defined. It is not easy to find a good consensus. In 1998 Duke University opened the "Center for Spirituality, Theology and Health," under the leadership of Dr. Harold Koenig. For the sake of practicality, I will use the definitions put forth by him in the book "Religion and Mental Health" [4]. Religion there is defined as "a set of beliefs, practices and rituals related to the transcendent, where the transcendent is that which relates to the mystical, supernatural or God in Western religious traditions, or to Brahman, Ultimate Truth, Ultimate Reality, or practices leading to Enlightenment, in Eastern traditions. (It) may also involve beliefs about spirits, angels or demons. Usually religion involves specific beliefs about life after death and rules to guide personal behaviors and interactions with others during this life. Religion is often organized and practiced within a community, but it can also be practiced alone and in private, outside of an institution, such as personal beliefs about and commitment to the transcendent and private activities such as prayer, meditation and scripture study." Regarding Spirituality Koenig says that "it is distinguished from its consequences (human values, moral, meaning, purpose, peace, connectedness, feelings of awe and wonder) by its link to the transcendent. The transcendent is that which is outside of the self, and yet also within the self, and in western traditions is called God, Allah, HaShem or a higher power and in Eastern traditions is called Ultimate Truth or Reality, Brahman, the Dharma or Budda."

R/S can play a positive or negative role when it comes to mental health. Among the positive, it may be associated with better coping, resilience, higher sense of meaning, and good mental health overall. Among the negative, religious beliefs may be incorporated in the psychopathology of the patient. About a third of schizophrenic patients suffer from religious delusions when psychotic. For bipolar disorder patients, the rate is 15 to 22% [5]. Religious hallucinations are also common.

Although traditionally Catholic, Hispanics with an interest in R/S issues have been open, more and more, to others' faiths and ways of believing when it comes to how to relate to the transcendent or sacred. It is not uncommon to find Hispanic patients that have embraced or frankly converted to non-Catholic Christian denominations ("Protestantism"), Buddhism, atheism (often times with an emphasis more on values and principles), Judaism, and others. It is not common (in my experience) to see Hispanic patients that convert to the Muslim faith.

Some of the main topics encountered in the literature about Hispanics, religion, and mental health include things like mental health stigma in Latinos, their use of religion for coping with stress, the possible protective role of religion in Latinos, the use of religion to increase access to mental health in Latinos and other minorities, etc.

In terms of protection from mental health problems, Meyers et al. looked at the data from the National Epidemiologic Survey on Alcohol and Related Conditions (NESARC) regarding religiosity, race, and alcohol use behaviors in the United States [6]. Frequency of religious attendance was inversely associated with alcohol use disorders. The association was more robust in Blacks than in Whites or Hispanics, but it also happened in the latter two groups.

Ai et al. investigated the association between childhood abuse, cultural factors, and lifetime substance use disorder (LT-SUD) among Latinas nationwide [7]. Religious attendance at a weekly level was negatively related to LT-SUD. The authors concluded that "religious attendance may have potential protection for Latinas."

Kirchner and Patino looked into religiosity in a sample of 295 Latino immigrants to Spain (63.1% of them were women) and found that religiosity was correlated with less depression, but only in women [8]. In women also, "the sense of spiritual fulfillment had mediating value in buffering the relationship between stress and depression symptoms." They conclude, naturally, that it is difficult to determine what part of the protection was due to religion and what part to social support received in the religious community.

Access to Mental Health Care in Latino immigrants is known to be significantly lower, compared with non-immigrant Latinos and non-Hispanic Whites. Faith Based Organizations (FBO) have been considered for helping this situation. According to the Department of Health and Human Services, one in four people with mental health concerns turn to faith leaders before they seek help from clinical professionals. Dalencour et al. report on 947 individuals from the Community Partners in Care Study, to see if they received depression care (determined by interview) in six FBOs in Los Angeles [9]. Sixty percent of the patients attended a religious place. Of those, 39% received depression services. The patients that used FOB mental health services had significantly higher mental health needs than those who did not use those services. This included prior diagnosis of depression, mania, or psychosis. Younger patients were more likely to use the services and to be African Americans and non-US born Hispanics. Due to the stigma associated with Mental Health in the Hispanic community, in general, many patients don't request or receive care. Frequently, in my experience, Hispanic patients are seen only when the situation can no longer be hidden like, for instance, in case of a suicide attempt, an episode of mania or psychosis, a panic attack that makes the patient go to the ER thinking they are having a heart attack, etc.

Another concern that Hispanic patients voice is that seeing a mental health provider (or just going to see a primary care provider), and having to register in a clinic, may come to haunt them back later as a problem with immigration services, if not with a quick detention on the spot followed by a speedy deportation.

Due the above two factors, efforts to "recruit" Hispanic patients into clinics that provide even virtually free care often times don't achieve much, and are met with frustration on the part of the staff in those clinics.

What could be some of the basic recommendations for providers that want to include R/S in their mental health practice?

1. Deal with your own issues first.
2. Elicit a thorough spiritual history.
3. Respect and support the patients' beliefs.
4. If necessary, challenge those beliefs.
5. Prayer with patients.
6. Consider working with a Chaplain.

1. I have heard Residents (and other practitioners) say that "we are not supposed to talk religion with our patients." Or that religion and politics should be left out of the therapeutic encounter. I agree that talking about politics is almost never helpful in psychiatric treatment. But R/S is different. Statements like the one from the World Psychiatric Association quoted above can do a lot to change that perspective. Residents and seasoned practitioners need to have a good understanding of how they perceive the themes of R/S, God, faith, and prayer. And to understand that patients have their own positions. Most importantly they need to understand that this areas need to be respected and explored regardless of our own beliefs or lack of.

   R/S also affect health providers. An interesting recent article published in the JAMA Journal of Psychiatry in May of 2020 found that religious attendance in almost 65.000 US nurses was associated with lower risk of death from despair [10].
2. Elicit a thorough spiritual history, including religious background and spiritual experiences throughout life (childhood, adolescence, adulthood). Determine how religion has been used to cope with life stressors. Any positive or negative experiences with religion should be recorded as well as religious beliefs and activities. As recommended by Koening, the response on the part of the patient that he is not spiritual should not stop a gentle exploration of the past. If the patient shows resistance to the topic, it is better to drop the subject. But especially when the patient engages in psychotherapy, the topic should be revisited at a later point, when the therapeutic alliance is more solid.
3. Respect and support the patients' beliefs. Respect, regardless of what they generate in you, in terms of countertransference, is important to the patient. And respect regardless of any degree of psychopathology you may perceive in those same beliefs. Take initially a more neutral exploratory stand, before deciding if they need to be supported or not. A great example of respect for religion in psychotherapy is in the book "Schopenhauer's Porcupine," by Deborah Anna Luepnitz, PhD. [11]. In Chap. 2 she describes the case of an 11-year-old Orthodox Jewish girl, daughter of a Rabbi, who presented with brittle diabetes (presumably due to psychological issues) that is beautifully and respectfully treated by a Catholic Psychologist.
4. If the beliefs are maladaptive, they need to be gently challenged. This is a touchy endeavor. It is better to do it once a therapeutic alliance has been established, and some progress in therapy in other areas has been achieved. A beautiful example of challenging spiritual beliefs is in the book "Love's Executioner" by Irvin Yalom, in the chapter "If rape was legal…" [12].
5. Prayer with patients. This is acceptable, if done with common sense and some basic requirements. The patient should initiate the request, the provider should have a good understanding of the patients' religious background and psychiatric diagnosis, both patient and provider should (ideally) share the same religious background, and the provider should have a clear understanding of what the prayer is for.

6. Consider working with a Chaplain. Any provider that believes that R/S are important issues when working in mental health but that does not feel well prepared or comfortable to deal with them, can refer the patient to a Chaplain, one that is ideally known to be able, in turn, to work with the provider (and not, for instance, recommend to the patient to stop taking psychiatric medication in order to let God heal him). Just like providers have specialists they feel more comfortable referring patients to for consultation, the same can be said of working with Chaplains.

Two personal vignettes may be worth mentioning briefly:

I was once called to see a 10-year-old Hispanic girl from Central America who came to the USA for treatment of osteosarcoma. The prognosis was grim and her mother (a highly educated person) was told from the beginning and in no uncertain terms. In fact, the girl died about a year later. The reason for the consult was that the mother expressed that if her only daughter was going to die despite receiving treatment she might as well "just kill her and then kill myself." I worked with that mother during that year. Two days before dying, the girl (who had dealt with the whole situation with nothing but grace, courage and no psychiatric issues) was telling me still that she "knew" that God was going to save her. For her mother, in psychotherapy for depression, the issue was not so simple. She was not particularly religious but she was open to the inclusion of the topic in therapy, at times. She felt challenged by her daughter's faith and found a degree of comfort in it. She was also able to speak about issues like feeling abandoned by God. In the end, she was able to accept her daughter's death with some sense of peace and she abandoned the idea of suicide. I followed her from a distance for at least four more years and she remained well adjusted to her loss and to life.

I was asked to see a 9-year-old boy from South America, who was battling a neuroblastoma. His case had a very poor prognosis. His father was a Physician and his mother a Psychologist. There were 3 more siblings involved, two older than the patient and one younger. I tried to work with this family as best as I could, but it felt up the hill for me, despite having excellent supervision. At some point I spoke with the father about their faith. He said they were Catholic. They said it was difficult to go to Mass because being with their son was a 24/7 duty plus Mass in English was "not the same." I offered to get them a Spanish speaking Priest, but the father declined, politely but firmly so I dropped the issue. The day the patient was dying, I stopped by to visit him. As I was arriving I noticed that he was totally alone in his room except for his mother, who was by the door, in a chair, quietly praying the Rosary. I spent some time with them. After a while, the father pulled me aside and told me that this was perhaps "the" moment to get that Priest if I could. I had recently become acquainted with a Priest from Spain that was spending some time in the city, and I went to his Church to tell him about the situation. He quickly went back with me to the Hospital and administered the last rites to my patient, with the whole family (and me), present. At the time I was agnostic, but I treasure the memory of that moment.

Conclusion

Religion and Spirituality need to be included in the psychotherapeutic encounter with Hispanic patients. The old antagonism between Psychiatry and Religion is weaning. For those interested in learning more about this, there are plenty of resources out there.

As we have discussed in this chapter, Psychiatry is changing its approach to R/S issues. Religion, in turn, is changing its approach to mental health. As an example, in a visit that American Psychiatrists paid to the late John Paul II, he told them:

"The meeting affords me a welcome opportunity to express the church's esteem of the many physicians and health care professionals involved in the important and delicate area of psychiatric medicine.... By its very nature your work often brings you to the threshold of human mystery. It involves sensitivity to the tangled workings of the human mind and heart, and openness to the ultimate concerns that give meaning to people's lives. These areas are of the utmost importance to the church, and they call to mind the urgent need for a constructive dialogue between science and religion for the sake of shedding greater light on the mystery of man in his fullness" [13].

References

1. https://news.gallup.com/poll/268205/americans-believe-god.aspx
2. Massachusetts General Hospital Handbook of General Hospital Psychiatry. 7th ed. Elsevier. 2018.
3. https://www.ncbi.nlm.nih.gov/pmc/articles/PMC4780301/
4. Koenig H. Religion and mental health. Research and clinical applications. Elsevier. 2018.
5. Koenig HG. Religion, spirituality, and psychotic disorders. Revista de Psiquiatria Clínica. 2007;34(1):40–8. https://doi.org/10.1590/S0101-60832007000700013.
6. Meyers JL, et al. Religiosity, race/ethnicity and alcohol use behaviors in the United States. Psychol Med. 2017 Jan;47(1):103–14.
7. Ai AL, Lee J. Childhood abuse, religious involvement, and lifetime substance use disorders among Latinas Nationwide. Subst Use Misuse. 2018 Nov 10;53(13):2099–111. https://doi.org/10.1080/10826084.2018.1455701.
8. Kirchner T, et al. Stress and depression in Latin American immigrants: the mediating role of religiosity. Eur Psychiatry. 2010 Dec;25(8):479–84. https://doi.org/10.1016/j.eurpsy.2010.04.003.
9. Dalencour M, et al. The role of faith-based organizations in the depression Care of African Americans and Hispanics in Los Angeles. J Psychiatr Serv. 2017 Apr 1;68(4):368–74. https://doi.org/10.1176/appi.ps.201500318.
10. Chen Y, et al. Religious service attendance and deaths related to drugs, alcohol and suicide among US healthcare professionals. JAMA Psychiat. 2020 Jul;77(7):1–9.
11. Luepnitz D. Schopenhauer's Porcupins: intimacy and its dilemmas. New York: Basic Books; 2002.
12. Yalom I. Love's executioner. NY: Harper Perennial Ed; 1990.
13. O'Connel R. The church and psychiatry. America: The Jesuit Review; 2001.

Further Reading

Sanchez AA. An examination of the folk healing practice of Curanderismo in the Hispanic community. J Community Health Nurs. 2018;35(3):148–61. https://doi.org/10.1080/07370016.2018.1475801.

Ai AL, Lee J. Childhood abuse, religious involvement, and lifetime substance use disorders among Latinas Nationwide. Subst Use Misuse. 2018;53(13):2099–111. https://doi.org/10.1080/10826084.2018.1455701.

Turner EA, Llamas JD. The role of therapy fears, ethnic identity, and spirituality on access to mental health treatment among Latino college students. Psychol Serv. 2017 Nov;14(4):524–30. https://doi.org/10.1037/ser0000146.

Dalencour M, Wong EC, Tang L, Dixon E, Lucas-Wright A, Wells K, Miranda J. The role of faith-based organizations in the depression Care of African Americans and Hispanics in Los Angeles. Psychiatr Serv. 2017 Apr 1;68(4):368–74. https://doi.org/10.1176/appi.ps.201500318.

Caplan S, Cordero C. Development of a faith-based mental health literacy program to improve treatment engagement among Caribbean Latinos in the northeastern United States of America. Int Q Community Health Educ. 2015;35(3):199–214. https://doi.org/10.1177/0272684X15581347.

Marques L, Eustis EH, Dixon L, Valentine SE, Borba CPC, Simon N, Kaysen D, Wiltsey-Stirman S. Delivering cognitive processing therapy in a community health setting: the influence of Latino culture and community violence on posttraumatic cognitions. Psychol Trauma. 2016 Jan;8(1):98–106. https://doi.org/10.1037/tra0000044.

Ransome Y, Haeny AM, McDowell YE, Jordan A. Religious involvement and racial disparities in opioid use disorder between 2004-2005 and 2012-2013: results from the National Epidemiologic Survey on alcohol and related conditions. Drug Alcohol Depend. 2019 Dec 1;205:107615. https://doi.org/10.1016/j.drugalcdep.2019.107615.

Dinh KT, Castro FG, Tein JY, Kim SY. Cultural predictors of physical and mental health status among Mexican American women: a mediation model. Am J Community Psychol. 2009 Mar;43(1-2):35–48. https://doi.org/10.1007/s10464-008-9221-9.

Hernandez M, Barrio C, Gaona L, Helu-Brown P, Hai A, Lim C. Hope and schizophrenia in the Latino family context. Community Ment Health J. 2019 Jan;55(1):42–50. https://doi.org/10.1007/s10597-018-0354-5.

Kirchner T, Patiño C. Stress and depression in Latin American immigrants: the mediating role of religiosity. Eur Psychiatry. 2010 Dec;25(8):479–84. https://doi.org/10.1016/j.eurpsy.2010.04.003.

Jurkowski JM, Kurlanska C, Ramos BM. Latino Women's spiritual beliefs related to health. Am J Health Promot. 2010 Sep-Oct;25(1):19–25. https://doi.org/10.4278/ajhp.080923-QUAL-211.

Caplan S. A pilot study of a novel method of measuring stigma about depression developed for Latinos in the faith-based setting. Community Ment Health J. 2016 Aug;52(6):701–9. https://doi.org/10.1007/s10597-016-0005-7.

Loue S, Sajatovic M. Spirituality, coping, and HIV risk and prevention in a sample of severely mentally ill Puerto Rican women. J Urban Health. 2006 Nov;83(6):1168–82. https://doi.org/10.1007/s11524-006-9130-x.

Baez A, Hernandez D. Complementary spiritual beliefs in the Latino community: the interface with psychotherapy. Am J Orthopsychiatry. 2001 Oct;71(4):408–15. https://doi.org/10.1037/0002-9432.71.4.408.

Krause N, Pargament KI, Hill PC, Ironson G. Assessing the role of race/ethnicity in the relationships among spiritual struggles, health, and Well-being. Am J Orthopsychiatry. 2018;88(2):132–41. https://doi.org/10.1037/ort0000255.

Villatoro AP, Dixon E, Mays VM. Faith-based organizations and the affordable care act: reducing Latino mental health care disparities. Psychol Serv. 2016 Feb;13(1):92–104. https://doi.org/10.1037/a0038515.

Ochoa CY, Haardörfer R, Escoffery C, Stein K, Alcaraz KI. Examining the role of social support and spirituality on the general health perceptions of Hispanic cancer survivors. Psychooncology. 2018 Sep;27(9):2189–97. https://doi.org/10.1002/pon.4795.

Sanchez AA. An examination of the folk healing practice of Curanderismo in the Hispanic community. J Community Health Nurs. 2018 Jul-Sep;35(3):148–61. https://doi.org/10.1080/07370016.2018.1475801.

A History of Mental Health Treatment in the Hispanic World

2

Fernando Espí Forcén

The Hispanic world entails a number of territories with the Spanish language as a common denominator. The medical and psychiatric care in these territories has had unique particularities linked to the different civilizations that invaded the Iberian Peninsula and the mix of ethnicities that populated the Americas and other territories throughout history. In this chapter, we review from a historical viewpoint how the care of the mentally ill has been shaped by these unique circumstances in Europe, America, and rest of the Hispanic world.

Iberia: Before "Hispania"

The area known today geographically as the Iberian Peninsula was named "Hispania" by the Romans after invading the area around the third century BCE. Under the Roman rule, the Greeks used the term Iberia to designate the peninsula below the Pyrenees. Before the Roman invasion, there had been in Iberia a few Greek colonies like Empúries, who participated in the Greek Olympics like any other Greek city. Also, the Phoenicians had come from the Middle East and had founded colonies in North Africa. One of these Phoenician colonies was Carthage that originated the Carthaginian civilization, a military empire in the Mediterranean Sea. The Carthaginians settled also in Southeast Iberia, founding the city of Qart Hadasht (Cartagena) and Sicily. Despite different colonizers the peninsula was populated by different people mainly the Celts who occupied two-thirds of the peninsula in the West and the so-called Iberians who populated one-third of the peninsula in the East. Among the Celts, there were Astures and Galacos in the northwest and Lusitanians in the west. The Celtiberians were Celtic people with influence from other Iberian people and lived in central Iberia. The Celts spoke Celtic language and had a Celtic

F. E. Forcén (✉)
Department of Psychiatry, Massachusetts General Hospital, Harvard Medical School, Chicago, IL, USA

R. Castilla-Puentes, T. Falcone (eds.), *Mental Health for Hispanic Communities*,
https://doi.org/10.1007/978-3-031-13195-0_2

culture. In Spanish, still there are some Celtic words that we have inherited from the Celtic culture such as "gancho" (hook) or "greñas" (long hair). The Iberians spoke different languages and included a variety of people including the Turdulos (former Tartessians), Turdetanos, Contestanos, and Edetani among others. Between the Celtic and the Iberian people lived the Vascones in central north who spoke Basque, a different language with Indo European roots. Basque language is still spoken today in this area [1, 2].

It is possible that the Celts and and Iberians inherited knowledge from Greek medicine as Greek and Phoenician traders visited frequently the peninsula, but the celts also had a medicine consisting of a mix between the magical and the use of plants for healing. Especially in Northwest Spain, there were different priests, being the druids the most valued. Druids used substances for healing like barley mixed with honey named "hidromiel" and liquor. They were considered the wisest people in their culture. Though Iberians had probably sophisticated medicine, to this day we have not yet been able to decode their writing [1, 2].

Hispania: The Greek-Roman Heritage

The Greek culture was assimilated by the Romans. With the invasion of the Iberian Peninsula by the Romans, the Celts and Iberians learned Latin and assimilated the Greek-Roman culture too. Iberia was called "Hispania" and became one province in the Roman empire. People living in Hispania gradually lost the Celtiberian languages and customs and became Romanized.

The philosophical and medical approach to mental illness in ancient Greece has been inherited to our present tense. For instance, Theophrastus of Eresus moved to Athens at a young age to study at Plato's Academy. Theophrastus described the different moral characters that people could have. Character was the Greek word for seal. For Theophrastus, character reflected on a predictable pattern of behavior that any person could have. Around the same time, Hippocrates was a Greek physician who lived in the island of Kos between the fifth and fourth century BCE. We don't know many details about the life of the Greek physician, but after his death the school of medicine was named after him. We know about his existence because he is mentioned in Plato's dialogues. The Hippocratic School understood all medical problems as an imbalance between the four humors in our body: the blood, the phlegm, the yellow bile, and the black bile. For ancient Greeks, the equilibrium of these humors would result in health or "eucrasia." In contrast, humoral disequilibrium would result in disease or "dyscrasia." Like other pathologies, mental illness was thought to result from dyscrasia. Thus, the Hippocratic School established a distinction between "mania" (due to problems with yellow bile), "melancholy" (in relation to black bile), and "frenesis" (a mania with fever). In the first century ACE, the Greek physician Dioscorides published the most popular pharmacopeia of the classic era. A century later, Galen of Pergamum embraced Hippocratic ideas and proposed more aggressive medical treatments such as purging and bloodletting in an attempt to restore eucrasia. Galen also discussed the four different temperaments that humans could have. As such, individuals could be sanguineous, phlegmatic, choleric, and melancholic. The four different temperaments set the base of all personality. Today psychologists differentiate between temperament as biologically based vs. character which is more environmentally influenced. After the fall of the Roman Empire, a group of Germanic people named the Visigoths from the area of Sweden invaded the Visigoths, invaded the Roman province of Hispania and created for the first time a judicial unity in the Iberian Peninsula. The Visigoths abandoned their Germanic languages and embraced Christianity and Latin. Due to the Visigoths, the most popular Spanish names are now Germanic like Alvaro, Gonzalo, Fernando, Carlos, Sancho, Lope, or Rodrigo. Also we have inherited Germanic vocabulary in the Spanish language from these Visigoths, like the words "guerra" or "tregua" [3–7].

Al-Andalus: Islamic Medicine in Iberia

Following the triumph of Christianity in the Western world, classic medicine fell in Europe. In contrast, Islam became the dominant religion in the East. Muslim doctors continued the legacy of Greek physicians at the Islamic hospitals named bimaristans or maristans (Persian word for hospital). In the eighth century, the Visigothic Hispania was conquered by the Umayyad caliphate and Hispania or Spania was renamed as Al-Andalus (land of the vandals). Eventually Cordoba became the capital of this Umayyad empire. In Baghdad, Avicena published in the eleventh century ACE the *Cannon of Medicine*. The book used an Aristotelian classification for the different medical afflictions. This book became the standard medical textbook in the Islamic empire. Some important physicians of Al-Andalus were the Sephardic physician Maimonides, father of medicine in the Jewish world, or the surgeon Albucasis among others. In Al-Andalus, the official language was Arab and people abandoned the Latin language. The Spanish language today entails a big amount of Arab words as part of its Muslim legacy [8–15].

From the Middle Ages to the Modern Era: The Creation of Hospitals for the Mentally Ill

Some Christian territories resisted the Muslim invasion and preserved their religion and the Latin language. They preserved the ideas of taking the Visigothic territories from the Umayyad empire under the Christian rule. They saw themselves as the direct descendants of these Germanic Visigoths that had once ruled the Iberian Peninsula. In these territories, a Romance language derived from Latin emerged between the ninth and eleventh centuries ACE. This Romance language was the origin of the Castilian. Other languages however emerged from Latin such as Asturian, Galician-Portuguese, Catalan, and Aragonese. Eventually Castilian became the dominant language among all these different territories. Because of that, Castilian is today also known as Spanish. This period of reconquering the Iberian Peninsula named La Reconquista took about 800 years and ended with the conquering of the Nazari kingdom of Granada by the Catholic monarchs in 1492. The same year Christopher Columbus arrived in the Island known today as the Dominican Republic. After the reconquering of Toledo by Alphonse VI of Castile, a number of humanists from all over Europe came to the city to translate the Islamic texts. Avicena's *Cannon of Medicine* was translated into Latin and became the most popular textbook in the Latin West until the mid-seventeenth century. However during the Middle Ages, due to lack of accessibility to Greek medicine, a spiritual approach was often taken to approach mental illness in the Hispanic kingdoms and rest of Europe [11].

During the Reconquista, a number of hospitals opened in Castile, Aragon, and Navarre (Hispanic kingdoms) in different territories to take care of outcasts, pilgrims, and the poor. In the end of the Middle Ages, the first hospital for specific treatment of the mentally ill was founded in the city of Valencia. One day in February of 1409, Father Joan Gilabert Jofre, a priest of the Mercedarian order was walking to the cathedral of Valencia to prepare his sermon when he saw a group of bullies beating up a man with mental illness in the street. Father Jofre stopped these people from harming this person. The following week he preached in the cathedral against the physical and sexual abuse that homeless people with severe mental illness experienced. Father Jofre proposed the construction of a hospital for the protection and care of people with mental illness. The sermon was heard by the bourgeois Lorenzo Salom who collected the funds for the construction of the new hospital. In 1410, the hospital of *Santa Maria d'Innocents, Folls e Orats* (Saint Mary of the Innocents, Mad and Demented) opened its doors. The name of the hospital, *Santa Maria d'Innocents, Folls e Orats* (Saint Mary of the Innocents, Mad and Demented), implies a classification of different mental pathologies. In Valencian language, *innocent* or *ignoscent* was a term used for people with intellectual disabilities. The Spanish translation would be *inocente* (in English innocent). *Foll* in Castilian Spanish could be translated as *loco* (in English mad or crazy). This term was likely used for people with mania or psychoses who would say "crazy things." *Orat* in Old Castilian Spanish could be translated as *dementado* (in English demented), a term used at the time for anyone deprived of reason or judgment [4, 16–20].

Following the foundation of the hospital in Valencia, the hospital of Holy Cross in Barcelona started to receive patients with mental illness in 1412. In 1419, the General Hospital of Jesus Christ in Cordoba also attended patients with mental health problems. In 1425, Alfonse V of Aragon, the new king of Aragon from the house of Trastamara, founded The Hospital of Our Lady Grace in Saragossa. This hospital soon started to welcome patients from all over the Iberian Peninsula (today Spain) and from different nations. In 1436, Marcos Sanchez Contreras founded the hospital of Saint Cosme and Saint Damien for the mentally ill in Seville. In 1445, the hospital of Saint Mary in Lerida also hosted patients with mental illness. The General Hospital of Palma de Mallorca also started to receive mentally ill patients in 1456. In 1483, the *Nuncio* (Vatican Diplomate) Francisco Ortiz travelled to Toledo and used his personal wealth to found the Hospital Nuncio in Toledo for the mentally ill. In 1489, the city of Valladolid also saw the foundation of a Hospital for the Innocents and Mad and in 1511 the Royal Hospital of Granada also received patients with mental illnesses. At the beginning of the modern era, the mentally ill in the Iberian Peninsula received standard medical care [4].

The diagnoses and treatments used in these hospitals were influenced by the classic Greek medicine that had reentered the Christian world through Islam. Therefore mania, melancholy, and frenesis were frequent diagnoses. The treatment also was focused on restoring the humoral balance. As such, purging, bloodletting, diet, and exercise were the common treatments. Chaplains would provide spiritual care and there were also workshops for occupational therapy and rehabilitations. The goal of the stay was always to cure the patient but those with chronic afflictions could reside in the hospital. If families could not pay, the cost of the stay was funded through charity. This model of medical care for the mentally ill was revolutionary in Medieval Europe and anticipated the Moral treatment of the Enlightenment [4, 16–25].

The Medical Approach in Pre-Columbian America

Before the arrival of the Spaniards to the new continent, the Aztecs, Mayans, and Inca people had developed relatively sophisticated medical treatments for their time. They had shamans and healers who used a combination of plants, dances, charms, and religious rituals with the goal of helping people with medical and mental afflictions restore health. In a magical way, shamans could inspect corn kernels or animal's guts to establish prognoses. They had obstetric experts to assist in deliveries and surgeons to cure and sew wounds. Therefore magical and somatic treatments were frequently intertwined. The use of psychedelic plants and fungus was also frequent in their rituals and healing practices. Psychedelic medicine has survived to this day. To date, psychedelic medicine is perhaps the biggest legacy of pre-Columbian medicine. Today psychedelics are widely considered in psychiatry for the treatment of depression, anxiety, and other mental afflictions. In addition, there is an underground psychedelic culture in society which entails trips to Peru, Mexico, and other Latin American countries in the search of shamanistic psychedelic treatments [26].

Medical and Mental Health Care in the Americas: The Foundation of the First Western Hospitals

The Greek classic model of medical care in the Iberian Peninsula was brought to America soon after the arrival of the first Spaniards. As such, already in 1503, Friar Nicolas Ovando founded in Santo Domingo the Hospital of Saint Nicolas of Bari. In 1524, and Hernan Cortes founded the Hospital of Jesus Nazarene in 1538 (Ayala-García, 2014).

Though all general hospitals regularly assisted people with mental illnesses, the first hospital for the treatment of people with mental illness in America was founded by Bernardino Alvarez in 1567. Bernardino was a former soldier who became a friar at the end of his life. After entering the clergy he volunteered at the hospitals of Mexico-Tenochtitlan. Seeing how there were not enough hospitals in the city to attend all patients who needed medical care, he used his personal wealth to found the hospital of Saint Hipolito. The local archbishop Alonso Montufar gave permissions for using the land next to the church of San Hipolito with the condition of using the same name. Bernardino proposed that patients deprived of reason should share time and space with those who had normal judgment as he believed virtue could be learned. Priests assisted with occupational activities and spiritual care. The first known physician to work in Saint Hipolito was José de Cabas. The treatment of the hospital was mainly European based. Thus, bathing and purging were often prescribed as remedies to treat mental illnesses. This hospital attended patients from all the different ethnicities at the time including people with mental illness of African, Native, and European origin. Saint Hipolito remained in use until 1910 [27–31]. The same model of Greek classical medicine was brought to the Philippines where the native treatments for medical problems were also used [32, 33].

The Contribution of African Medicine in the Americas

Despite the existence of hospitals, there was in general a scarcity of European physicians in the Americas. Many African people were brought by Europeans to work as slaves in the new continent. As part of the slave trade, different African plants in the form of seeds came as well. Some of these plants were used for healing purposes. A number of magical healing rituals with the use of plants emerged in the Americas concomitantly with European and Native American medicine. These African medical practices persist to this day. Voudon or Voodoo, Santeria, Gullah, Shango, and Tambor das Minas are examples of these kinds of practice. African priests and healers found in the Americas some of the same plants they had used in Africa. At the same time they learned how to use Native American plants with the help of Spanish Jesuit Priests [34].

The Crisis in the Nineteenth Century and the Birth of Modern Neuroscience

After Napoleon invaded Spain and the beginning of the peninsular war, Francisco de Miranda, Simón Bolívar, Antonio José de Sucre, José de San Martín among others declared the independence from the Iberian Peninsula leading to the establishment of a number of independent republics in the Americas. During the peninsular war, Napoleon troops bombed the hospital of Saragossa, at the time the most important psychiatric hospital in Spain. Many patients with mental illness died during the bombing. Father Monal, a catholic priest, found refuge for those who survived the bombing. The quality of medical care for the mentally ill declined in the Iberian Peninsula following the French invasion. Following the Spanish independence from France, the new government ordered ecclesiastical confiscations. The problem is that the hospitals for the mentally ill became organized by the public local administrations who lacked resources. As a result, the care for the mentally ill declined. In both Spain and Latin America, the nineteenth century was dominated by frequent and disastrous fraternal civil wars. The golden years of psychiatric and medical care in the Hispanic world came to an end. Nonetheless, the new ideas that came from North Europe like phrenology and mesmerism arrived in Spain and Latin America too. It is important to note that despite the health crisis in the Hispanic world some scientists became international leaders. Perhaps the most significant example is the figure of Santiago Ramón y Cajal who probed the neuron theory, described the dendritic spines and the direction in which neurons communicate with each other, developed a model for memory formation and anticipated neuroplasticity. In the early twentieth century, he was awarded the Nobel Prize in Medicine together with Camilo Golgi. As such, Cajal can be considered as the father of modern neuroscience and a prominent historical scientific figure in the Hispanic world [4, 19, 35].

References

1. Lorrio AJ. Los Celtíberos. Universidad Complutense de Madrid; 1997.
2. Dietler M, Lopez-Ruiz C. Colonial encounters in ancient Iberia: Phoenician, Greek, and indigenous relations. Chicago: University of Chicago Press; 2009.
3. Kleisiaris CF, Sfakianakis C, Papathanasiou IV. Health practice in ancient Greece: the Hippocratic ideal. J Med Ethics Hist Med. 2014;7:6.
4. Gonzalez Duro E. Historia de la locura en España. Madrid: Temas de Hoy; 1996.
5. Porter R. Madness: a brief history. New York: Oxford University Press; 2003.
6. Shorter E. A history of psychiatry. New York: John Wiley and Sons; 1997.
7. Barcia D. Historia de la psiquiatria española. Madrid: You and Us; 1996.
8. Wenner MW. The Arab/Muslim presence in medieval Central Europe. Int J Middle East Stud. 1980;12(1):59–79.
9. Al-Djazairi SE. The hidden debt to Islamic civilisation. Manchester: Bayt Al-Hikma Press; 2005.
10. Collins R. The Arab conquest of Spain, 710–797. Oxford: Blackwell; 1989.
11. Roth N. Jews, Visigoths and Muslims in medieval Spain: cooperation and conflict. Leiden: Brill; 1994.
12. Dols MW. The origins of the Islamic hospital: myth and reality. Bull Hist Med. 1987;61(3):367–90.
13. Dols MW, Immisch DE. Galen and mental illness. In: Dols MW, editor. Majnūn: the madman in medieval Islamic society. Gloucestershire: Clarendon Press; 1992.
14. Menocal MR. The ornament of the world: how Muslims, Jews and Christians created a culture of tolerance in medieval Spain. Boston and New York: Back Bay Books; 2002.
15. Girón F. Oriente islámico medieval. Madrid: Akal; 1994.
16. Hinojosa JR. Hospital General de Valencia 1512–2012. Valencia: Fundació Hospital Reial i General; 2012.
17. López-Ibor JJ. La Fundación en Valencia del Primer Hospital Psiquiátrico del Mundo. Actas Esp Psiquiatr. 2008;36(1):1–9.
18. Espí Forcén F. A hospital for the mentally ill in the middle ages-psychiatry in history. Br J Psychiatry. 2016;208(2):103.
19. Espí Forcén F. Casas de locos: the treatment of the mentally ill in the Hispanic world before the enlightenment. J Nerv Ment Dis. 2008;207(9):1.

20. Merenciano FM. Vida y Obra del Padre Jofre. Fundador del primer Manicomio del Mundo, 1409. Madrid: CESIC; 1950.
21. Garcia M, et al. La locura en el hospital de los inocentes de Sevilla (1436–1840). edupsykhé. 2014;13(1):5–12.
22. Perez J, Baldessarini RJ, Udurraga J, Sanchez-Moreno J. Origins of psychiatric hospitalization in medieval Spain. The Psychiatric Quarterly. 2012;83(4):419–30.
23. Sancho de San Roman R. El hospital del Nuncio de Toledo en la historia de la asistencia psiquiátrica. Anales Toledanos. 1983;17:55–72.
24. de Heredia-Torres P, Brea-Rivero M, Martinez-Piedrola RM. The origins of occupational therapy in Spain. Revista de Neurologia. 2007;45(11):695–8.
25. Espinosa Iborra J. Un testimonio de la influencia de la psiquiatría española de la ilustración en la obra de Pinel: el informe de José Iberti acerca de la asistencia en el Manicomio de Zaragoza (1791)., Asclepio; 1964.
26. Carod-Artal FJ. Hallucinogenic drugs in pre-Columbian Mesoamerican cultures Alucinógenos en las culturas precolombinas mesoamericanas. Neurologia. 2015;30(1):42–9.
27. Ayala-Garcia MA. The first 10 hospitals in the American continent. Chicago: Hektoen International; 2014.
28. Díaz de Arce J. Libro de la Vida del Proximo Evangelico el Venerado Padre Bernardino de Álvarez. Mexico: Digital Copy Obtained from Universidad Autónoma de Nuevo León; 1761. http://cdigital.dgb.uanl.mx/la/1080026339/1080026339_MA.PDF
29. Quijano-Pitman F. El Primer Hospital Psiquiátrico de América, 1567, Primicias Médicas Nacionales. Gac Méd Méx. 1997;133(4):351.
30. Leiby JS. San Hipólito's Treatment of the Mentally Ill in Mexico City, 1589–1650. The Historian. 1992;54(3):491–8.
31. Espí Forcén F. Bernardino Alvarez and the origins of psychiatric care in America. Br J Psychiatry. 2019;214(02):89.
32. Bantug JP. A short history of medicine in the Philippines during the Spanish Regime 1565–1898. Manila: Colegio Médico-Farmacéutico de Filipinas; 1953.
33. Clain P. In: Correa J, editor. Remedios fáciles para diferentes enfermedades. Manila: En Manila en el Collegio y Universidad de Sancto Thomas de Aquino; 1712.
34. Voeks R. African medicine and magic in the Americas. Geographical Rev. 1993;83(1):66–78.
35. DeFelipe J. Sesquicentenary of the birthday of Santiago Ramón y Cajal, the father of modern neuroscience. Trends Neurosci. 2002 Sep;25(9):481–4.

Mental Health Stigma in the Latinx Population: Treatment Implications and Future Directions

3

Stacie Craft DeFreitas

Introduction

Mental health stigma, also known as mental illness stigma, is the social judgement or devaluation of a person with mental illness or of mental health treatment. Because those with mental illness may experience discrimination such as denial of housing and employment, there are real-life consequences for those who face stigma. Often, such stigma beliefs are driven by lack of knowledge about mental health, but among Latinx communities, mental health stigma may have a particularly profound impact as a result of their cultural values and norms as well as experience with racial discrimination. This chapter examines mental health stigma in the Latinx community by highlighting the various types of stigma, the potential negative outcomes of stigma, factors that influence rates of stigma, and how practitioners can address mental health stigma.

Types of mental health stigma:

Mental health stigma is complex in its conceptualization, but for the purposes of this chapter, the focus will be on perceived- and self-stigma. Perceived (public) stigma concerns beliefs that others have negative views about those with mental illness or about mental health treatment. Public stigma is often assessed using measures of social distance or by asking stigma-related questions about vignettes concerning individuals with mental health problems. Perceived stigma beliefs can cause people to fear being labeled as a person with mental illness because of concerns about being judged negatively. Latinx parents have reported concerns that their family, friends, and community members would stigmatize them if they sought treatment for their children who suffered from worry and sadness [1]. Not only can perceived stigma beliefs impact treatment seeking, they can also impact mental health. For example, youth, including Latinxs, with higher rates of perceived stigma

S. C. DeFreitas (✉)
Department of Psychology, Prairie View A & M University, Prairie View, Texas, USA
e-mail: scdefreitas@pvamu.edu

R. Castilla-Puentes, T. Falcone (eds.), *Mental Health for Hispanic Communities*,
https://doi.org/10.1007/978-3-031-13195-0_3

had greater odds of suicidal ideation and attempts [2]. Perceptions that others harbor stigma about mental illness and treatment impair treatment seeking and mental health overall.

Self-stigma, which is also known as personal stigma, is when public stigma becomes internalized such that individuals with mental health problems have negative beliefs about themselves due to their illness. Such beliefs can result in lower self-esteem and are related to feelings of shame [3]. In a predominantly Latinx sample of youth under psychiatric treatment, all but one of them reported feelings of shame related to their diagnosis that led to feelings of low self-worth [4]. Perceived stigma and self-stigma are related such that believing that others have mental health stigma beliefs is related to greater self-stigma [5]. Perceived- and self-stigma, both have deleterious impacts on mental health care.

Other types of mental health stigma are also important to consider, though they have been less widely researched. Structural stigma focuses on the institutional consequences of mental health stigma. An example of this would be that Latinx Americans are more likely to believe that those with mental illness are dangerous than European Americans [6] and such beliefs can have institutional consequences such that a person with mental illness would potentially be denied housing and jobs because of this false belief about dangerousness. Another structural stigma issue is that Latinx parents are concerned about such matters for their children, noting worry about whether their children will be placed in special education or have difficulty obtaining employment as a result of seeking treatment [7]. They also have concerns about losing their parental rights [1]. Another type of stigma is intersectional stigma which involves the combination of multiple stigma identities within a person or group and is sometimes known as double stigma. The term double stigma can be misleading however, as it suggests that the effects of stigma are merely added together when in truth they interact in a much more complicated, multiplicative manner [8]. Intersectional stigma is affected by how severely each of the identities differs from the social norm and how much victim blaming is associated with each identity. Therefore a gay Latinx man with a substance abuse problem would likely experience more severe stigma than a heterosexual Latinx woman with an anxiety disorder. It is important to remember that mental health stigma is found among all ethnic/racial groups within the United States. This is not a problem that is unique to the Latinx community, but understandings of mental health stigma that consider culture are important to addressing this problem.

Levels of Mental Health Stigma in the Latinx Community:

Various studies have been conducted comparing rates of mental health stigma between sectors of the Latinx community and other ethnic/racial groups with mixed results (Table 3.1). Some have found Latinxs to have higher rates of mental health stigma than other groups such as European Americans [7, 9, 10]. Others have found them to exhibit lower rates of mental health stigma than African Americans [11, 12] and Asian Americans [5]. In contrast, studies have found that the stigma beliefs of Latinx populations are similar to European Americans [12] and African Americans [13]. Comparison studies present an unclear picture and may not be most helpful in developing understanding of the impact of mental health stigma in the Latinx population.

Perhaps more important are studies that only focus on Latino populations when examining rates of mental health stigma (Table 3.2). Studies that examine the

Table 3.1 Ethnic comparisons of mental health stigma

Sample	Findings	Source
609 college students of color	Asian American students exhibited higher rates of stigma than Latinx or African American students	Cheng et al. (2013)
238 parents of children in Head Start, 17% Latinx	Latinx parents had less mental health stigma beliefs than African American parents, but were similar to European American parents	Turner et al. (2015)
29 families with children with anxiety	70% of Latinx parents reported stigma-related concerns about seeking treatment while 17% of European American parents reported concerns	Chavira et al. (2017)
122 African American and Latinx college students	Latinx college students exhibited lower rates of self-stigma and perceived stigma than African American students	DeFreitas et al. (2018)
271 participants from a primary care clinic	Latinx participants had higher self- and perceived stigma than European American participants	Benuto et al. (2019)
55 adolescents of color	African American and Latinx adolescents did not differ in mental health stigma	Wang et al. (2019)
667 adolescents	Latinx and African American youth had greater desire for social distance from peers with mental illness than European Americans	Dupont-Reyes et al. (2020)

Table 3.2 Rates of mental health stigma in Latinx samples

Population	Rates of MHS endorsement	Study
Depressed community sample	51% endorsed at least one MHS belief	Vega et al. (2010)
Latinx parents of children with anxiety	70% reported stigma concerns about seeking treatment for their children	Chavira et al. (2017)
Depressed, Spanish speaking Latinx	83% of sample endorsed at least one MHS belief	Collado et al. (2019)
Adolescents, 45.5% Latinx	86.4% of the youth cited stigma as a barrier to seeking treatment	Wang et al. (2019)

levels of mental health stigma in the Latinx community find that it is typically present in the majority of the sample. For example, among depressed, Spanish speaking Latinxs, 83% endorsed at least one mental health stigma belief [14]. When examining adolescents, 86.4% of a diverse students sample reported that stigma was a barrier to seeking treatment [13]. It is important to note that mental health stigma rates may be even higher than is suggested by research due to factors such as social desirability. Though individuals may endorse low rates of stigma on a survey, they may really have greater stigma beliefs. This is suggested by a study conducted with a primarily Dominican sample using a method called Preguntas con Cartas (Questions with Cards) [15]. Participants were presented with questions about mental health stigma, then permitted to respond using cards that were placed face down and compiled anonymously in the group. The anonymous group score indicated higher levels of stigma when compared with survey responses about stigma. This discrepancy suggests that the ability to respond anonymously reduced pressure to answer in the socially desirable manner. These findings indicate that mental health stigma is a concern in the Latinx community that should be combatted.

Impact of Mental Health Stigma on Treatment

Mental health is particularly important because of its impact on treatment. For those who harbor self-stigma, they may be reluctant to even seek the knowledge that they need in order to address their mental health needs. The shame and fear of someone finding out about their issues and being labeled as mentally ill may be too great. Latinx individuals are less likely than European Americans to seek mental health treatment [9]. Though there are a variety of external barriers to treatment such as less access to mental health insurance, even when studies control for these factors, the discrepancy remains. For example, Benuto and colleagues [9] found that even when controlling for age, gender, income, and education among a community sample, 70.3% of European Americans and 34.6% of Latinxs reported using mental health treatment. Mental health stigma likely plays a significant role in this difference.

Mental health stigma beliefs are likely related to a lower desire to seek treatment among Latinx populations (Table 3.3). This appears to be the case in community

Table 3.3 Relationships between mental health stigma to treatment and mental health outcomes

Stigma Construct	Outcome	Population	Study
Self-stigma	Greater stigma predicted lower mental health service use	Community sample	Benuto et al., (2019)
	Greater stigma predicted lower willingness to seek mental health service	Mexican American college female students	Choi et al. (2019)
Perceived stigma	Higher perceived stigma was related to less likelihood to disclose mental health issues to family/friends, less likely to take depression medication and more likely to miss mental health appointments	Community sample with depression risk	Vega et al. (2010)
	Lower stigma was related to more positive attitudes about seeking help	Latinx college students	Mendoza et al. (2015)
	Parent stigma was related to a lower intention of seeking help for their children	Latinx parents of children in Head Start	Turner et al. (2015)
	Parent stigma was related to a greater belief that they would delay or not seek treatment for their children	Latinx parents of third and fourth graders	Young et al. (2015)
	Greater stigma was related to higher odds of suicidal ideation and attempts	A diverse sample of students of color that included 8402 Latinxs	Goodwill, Zhou (2020)
General MHS	Parents with stigma were less likely to seek treatment for their anxious children	Parents of children with anxiety, 34.5% Latinx	Chavira et al. (2017)
Stigma related to eating disorders	Higher rates of stigma were related to lower likelihood of seeking or receiving treatment	Latinx females with a history of bulimia or binge eating disorder	Higgins et al. (2019)

samples [9, 16], among parents [12], and within college samples [17]. Perceived stigma beliefs were related to being more likely to miss scheduled mental health meetings and being less likely to take antidepressants among a community sample [16]. In line with this finding, having lower rates of public stigma is related to more positive attitudes about help-seeking [18]. When Latinx parents have concerns about mental health stigma, they are less likely to seek treatment for their children who suffer from anxiety [7]. Specifically, concerns that their child's teacher will find out about their mental health treatment, that this knowledge would reflect poorly on them as parents, that the child would be teased, and that the parent would be embarrassed were all related to avoiding seeking treatment or delaying seeking treatment for their children [19]. Similar results are also found in college students. Mexican American college students with higher stigma concerning mental illness being incurable and beliefs that those with them were dangerous and untrustworthy were more like to have a preference for no treatment or for religious treatment as opposed to conventional treatment [20]. Further, among Latinx women with a history of eating disorders, having stigma about eating disorders was related to a lower likelihood of them seeking or participating in treatment [21]. Despite this collection of findings, it is important to note that mental health stigma does not always predict service use for those in the Latinx community. A meta-analysis, that included two studies, found that among Latino, Cuban, and Puerto Rican samples, mental health stigma was not related to help-seeking [22]. Further, Sánchez and colleagues' [23] examination of a Puerto Rican, immigrant Latino, and Brazilian sample also found no relationship between stigma and help-seeking; however, it is important to note that their stigma measure focused on beliefs about whether people with mental illness are discriminated against. Despite these few findings that do not demonstrate a relationship, there is significant evidence to suggest that among Latinx communities, having mental health stigma reduces the likelihood that individuals are willing to seek mental health treatment.

Factors that May Increase Mental Health Stigma

In order to challenge the impact of mental health stigma in the Latinx community, factors that potentially foster it must be examined (Table 3.4). Many mediators of the relationship between mental health stigma and treatment outcomes have been examined, but those that seem to have the most profound impact include past treatment experiences, perceived racial discrimination, mental health knowledge, and cultural values.

Treatment can have a positive or negative impact on mental health stigma. Latinx clients, particularly immigrants, often have problems with dropping out of treatment early or with receiving care that is of lower quality. When treatment experiences are negative, such experiences are likely to increase mental health stigma and may in turn reduce further mental health service use [24]. However, when experiences are positive, they may reduce mental health stigma. For example among depressed, Spanish speaking clients receiving supportive counseling in which they

Table 3.4 Factors related to higher mental health stigma

Construct	Potential Impact	Source
Acculturation	No impact on mental health stigma	Choi et al. (2019)
	Higher levels of acculturation were related to fewer negative beliefs about mental illness and therapy	Rojas-Vilches et al. (2011)
Age	Parents had greater social stigma concerning mental health and a greater belief that mental illness was incurable than their college-aged children	Rojas-Vilches et al. (2011)
College education	Those with more education had greater stigma concerning taking antidepressants and less desire for social distancing	Lopez et al. (2018)
Depression	Higher rates of depression were related to higher rates of stigma	Collado et al. (2019)
Enculturation	Those with higher enculturation had greater stigma	Hirai et al. (2015)
Mental health knowledge	Those with greater mental health knowledge exhibited less mental health stigma, but more stigma about antidepressants	Lopez et al. (2018)
Social anxiety	Having anxiety about interacting with someone with a mental health disorder was related to greater mental health stigma	DeFreitas et al. (2018)
Spanish media preference	Having a preference for Spanish media over English media was related to greater social distance preference	Dupont-Reyes et al. (2019)

were permitted to vent their concerns, mental health stigma decreased over time and more so than those in behavioral activation treatment [14]. The authors suggested that the venting process may be compatible with clients' perceptions of what therapy is and therefore reduce stigma. They suggest allowing clients to vent even within therapy formats that do not typically allow for that process. Positive therapy experiences reduce negative beliefs that individuals have about therapy, allowing them to reap the benefits of therapy.

When individuals have experiences with racism, these negative interactions can also impact their levels of mental health stigma. For example, among a diverse sample of college students, the more discrimination that they experienced, the more likely they were to anticipate greater perceived stigma concerning mental health help-seeking and in turn had greater self-stigma [5]. Quinn and colleagues [25] suggest that when a person has experiences with racial discrimination, then this increases their belief in the likelihood that they would be discriminated against as a result of a concealed stigmatized identity such as having a mental illness or substance abuse problem. Further, racist stereotypes such as Latinx individuals being prone to drug abuse, may be related to increased stigma for psychotropic medications such as antidepressants when compared to psychotherapy [26]. Beliefs that antidepressants are addictive [27] create hesitancy in the Latinx community in which they do not want to associate with drugs, even medications to treat mental illness. Experience with racism may foster greater stigma towards medication to treat mental illness and make individuals more likely to expect discrimination due to having a mental illness.

Another factor that impacts mental health stigma is knowledge about mental health or mental health literacy. For example, some Latinx immigrants harbor stigma beliefs about depression because they believe that it is caused by evil forces and sinfulness [28]. A similar conclusion was found from a study among Mexican American college women who were more likely to harbor self-stigma beliefs if they believed that the causes of mental illness were spiritual in nature (i.e., punishment for sins) [29]. When individuals have better understanding about the biological etiology of mental health disorders, they have fewer stigma views [30]. Knowledge may also be important because increasing knowledge is likely to reduce the level of anxiety someone has about interacting with someone with a disorder as they will have a better understanding of what to expect. Having less anxiety about social interactions with mentally ill individuals is related to less mental health stigma [11]. With psychoeducation about how mental illnesses present and treatment options, mental health literacy can be increased and stigma, in turn, reduced.

One of the most important factors that one must consider when examining mental health stigma in the Latinx community is the influence of cultural values and norms. One compelling study found that a preference for consuming media in Spanish versus English is related to higher levels of mental health stigma among youth [31]. The assumption is that Spanish media (film, television, radio, or music) is less likely to include anti-stigma messages and information about mental health help-seeking, perhaps as a result of cultural beliefs. Further evidence that connects Latinx culture to stigma was found in a study of Mexican American college students in which those with greater enculturation—connection to Mexican culture as opposed to mainstream American culture—was related to higher levels of mental health stigma as well as a greater preference for no treatment or religious treatment as opposed to conventional treatment [20]. One's level of acculturation may also impact mental health stigma, but research has been mixed. Studies of Mexican American female college students found no relationship between acculturation and mental health stigma [29] while acculturation was related to less stigma about mental illness and treatment among Latinx college students and their parents [17]. Overall, these examinations of connection to Latinx culture in general suggest that there may be some aspects of the culture that foster mental health stigma.

Considerations of more specific aspects of Latinx culture also demonstrate a connection to mental health stigma. Family or courtesy stigma refers to the belief that family and friends may experience discrimination as a result of their relationship to a person with a mental illness [3]. Though not examined well, family stigma is likely important for Latinos due to familismo beliefs. Familismo refers to the high value in family connection within many Latinx groups. Family members may be blamed for their kin's mental illness or others might perceive that they could be "infected" by it. Because of stigma beliefs, despite their strong connection to family, many members of the Latinx community are hesitant to share their mental health concerns with their family members. Among Latinx at risk for depression, greater perceived stigma was related to less likelihood that they would disclose their depression diagnosis to family or friends [16]. Research has also connected familismo to treatment use finding that those with high familismo behaviors were more likely to utilize informal or religious mental health treatment than medical or specialty psychological interventions [32]. These findings indicate that valuing family may complicate the treatment process of members of Latinx communities because of stigma beliefs.

The hesitancy to disclose mental health problems to others is likely due to the judgement that individuals anticipate from sharing that they are having mental health difficulties. This may be particularly problematic among Latinx youth who report greater desire for social distance from those who have mental health problems than European American youth [10]. Latinx adolescents, particularly males and those with higher rates of public stigma, likely expect that disclosure will result in social isolation. When such isolation includes the family, this may worsen mental health outcomes, so many choose to remain silent.

Other Latinx cultural ideas such as machismo and marianismo may also impact mental health stigma. Marianismo suggests that Latinx women should be dignified and always self-sacrifice even during suffering while machismo values state that Latinx men should be strong and the protectors of their families. These cultural beliefs may influence mental health stigma because having a mental illness would be seen as a sign of being weak (débil) as it goes against these cultural views. Further some members of the Latinx community feel as if they are not worthy of respect if they must seek treatment [33]. The belief that one is not fulfilling cultural norms and expectations if they seek treatment can have a profound impact on one's judgement of mental health and treatment. These beliefs can influence feelings of self-worth making denial that one has a problem and avoidance of treatment a likely outcome.

Interventions to Reduce Mental Health Stigma

Due to the profound impact that mental health stigma can have on the lives of those with mental illness, various intervention programs have been developed to reduce such beliefs. Unfortunately, many of these programs fail to consider culture or even include Latinx participants [34]. Contact programs are interventions in which individuals are exposed to people who have mental health disorders. These interventions can be done in person, using video, or online, but in vivo approaches have a more significant impact on reducing mental health stigma [3]. Among adults, contact programs typically lead to reductions in mental health stigma and the willingness to seek treatment in turn [34]. However, among adolescents, such programs do not have this effect. One study of adolescents in which 40% of the sample was Latinx found that contact with college students who had a history of bipolar disorder did not result in reductions of mental health stigma [35]. This was similar to Corrigan and colleagues' [34] finding from their meta-analysis that contact interventions are ineffective with adolescent populations. Contact programs may reduce mental health stigma among Latinx adults, but are likely not useful for adolescents; however, little has been done to examine whether culture is an important consideration for such interventions.

One culturally adapted intervention method that has been used to reduce mental health stigma among Latinxs is the use of an educational fotonovela. A fotonovela is an entertainment-education strategy that uses a popular media form to increase knowledge, shift social norms, and induce individuals to engage in socially responsible action. Typically they include cultural norms, popular images, simple language, and colorful, entertaining drawings that are designed to be engaging. Ideally the characters are perceived as similar to the reader, likeable and real. One such tool is "Secret Feelings." Cabassa and colleagues [36] developed this depression focused fotonovela with input from primarily Mexican and Central American individuals. It was designed to help readers know that depression is a common disorder that, though it is serious, can be treated. "Secret Feelings" challenges common misperceptions about the addictiveness of depression medications and that depression is related to weakness, as well as encourages speaking to a doctor about symptoms.

"Secret Feelings" has been used to reduce mental health stigma in a variety of Latinx samples (Table 3.5). A randomized clinical trial using a convenience sample of adult night school students found that the use of the fotonovela did not reduce desires for social distancing or perceptions of dangerous more than the control group, though it did result in greater knowledge about depression treatment [37]. The authors did note that overall the sample had very low social distancing scores and this floor effect could account for the lack of findings. Another study demonstrated that "Secret Feelings" can reduce social distancing desires from pre- to post-test among a mostly Spanish speaking sample of Latinx women with depression [38]. Among a sample of immigrant Latinx women, those who read "Secret Feelings" had greater decreases in antidepressant stigma, but no change in general mental health stigma when compared to a control group [39]. Their exposure to the fotonovela also increased their depression knowledge and intent to seek treatment. One important aspect of the fotonovela is that it is designed to be shared with family and friends as it is entertaining and generates conversation. Among Latinx participants from a community adult school, 55% of those who received the fotonovela reported passing it on to someone else compared to 36% of those who were given a depression pamphlet [40]. This finding suggests that the fotonovela may be a good tool to spread

Table 3.5 Summary of fotonovela, "Secret Feelings," and mental health stigma studies

Study	Sample	Outcomes
Hernandez et al. (2013)	142 immigrant women, primarily from Mexico	Fotonovela group had greater decreases in antidepressant stigma but not change in stigma about mental health treatment generally than control group
Unger et al. (2013)	157 adults in a community ESL school	Exposure to the fotonovela was related to greater reduction in antidepressant stigma and mental health stigma when compared with those given a depression pamphlet
Cabassa et al. (2015)	132 adult night school participants, primarily Mexican	Fotonovela group did not differ from depression brochure group on social distance or perceptions of dangerousness
Sanchez et al. (2019)	305 primarily Spanish speaking and female Latinx with depression	Exposure to the fotonovela was related to less desire for social distance from those with mental illness and less stigma about general mental health care, but increases in antidepressant stigma

information about mental illness throughout the Latinx community. Overall, "Secret Feelings" appears to be a culturally appropriate tool for some Latinx individuals and is a good resource for primary care and mental health care providers.

Ways to Address Mental Health Stigma in Treatment

Mental health care providers should strive to assess stigma beliefs and address them within the therapy process (Table 3.6). Basic tasks such as psychoeducation and focusing on building rapport that are typical to the therapy process are critical to

Table 3.6 Mental health practitioner strategies to address stigma concerns

Specific Latinx Population	Strategy to address mental health stigma
General	Assess stigma at intake and discuss stigma concerns during the first session
	Explore possible mistrust in the therapy relationship and address any concerns
	Avoid a color-blind mentality and address issues of ethnicity and race, particularly if the provider is not ethnically matched with the client
	Allow clients time in therapy to vent their frustrations
Depressed clients	Because of high levels of antidepressant stigma, assess stigma beliefs in this area. Reserve discussions of antidepressants until rapport has been built and only for those with severe depression.
Familismo believers	Inclusion of family in the therapy process for psychoeducation and to build a support network for the client.
K-12	Treatment in the schools should be conducted in a discreet manner in order to address stigma concerns including making sure that the location of treatment support confidentiality
LGBTQ+	Consider how the additional stigma due to sexual orientation may result in a more complicated intersectional stigma
Mothers	Consider that mothers may be experiencing additional stigma if they are having difficulty caring for their children which should be addressed in therapy.
Parents with children with mental health needs	Address concerns that parents may have, including the feeling that their child's mental health problems reflect their poor parenting
Psychotic	Recognize that they may have particularly strong feelings of self-stigma that impact their relationships with family and friends that need to addressed
Religious	Collaborate with trusted religious healer in the community to incorporate religious values into the therapy practice
Spanish speaking preference	Survey all clients' language preference and discuss what it means if it cannot be fulfilled. Provide options for referral if necessary.
	Distribute mental health information in Spanish, perhaps using a culturally appropriate tool such as a fotonovela
Undocumented	Consider the additional stress concerning their immigration status, potential trauma from crossing, work conditions, and how they can potentially impact mental health stigma

reducing mental health stigma in current patients; however, it is critical to assess the individual needs of each client and consider culture when treating Latinx clients.

Increasing the Number of Culturally Matched Providers

Though Latinx Americans account for almost 20% of the population in the United States, only 4% of the mental health psychologists are Latinx [41]. A greater number of Latinx mental health providers are needed to assist in the struggle to reduce mental health stigma and provide effective therapy in the Latinx community. Latinx mental health providers typically have the cultural knowledge to develop treatments and interventions that can address stigma concerns and reduce mental illness symptoms. However, this does not mean that members of other ethnic groups cannot provide effective treatment to Latinx individuals. With cultural competence, appropriate and efficacious treatment can be produced.

Mental health practitioners have a responsibility to address issues of culture in the therapy process. Principle E of the APA Ethics code states that psychologists must consider and respect cultural differences [42]. There are a variety of cultural factors that should be considered when treating members of the Latinx community including personalismo. The cultural value of personalismo—valuing warm and emotional connections—suggests that the relationship between the client and the mental health provider may be even more important than the provider's expertise. Building a strong, trusting relationship should be the first priority of the therapist.

A key idea noted by Abdullah and Brown [33] is that interventions should not be designed to change cultural values, but instead to emphasize values that support less stigma or shifting the client's views of mental illness or treatment so that it fits better with their cultural values. For example, instead of trying to change machismo values, demonstrate how engaging in therapy will allow a severely depressed father to be the protector that he desires because he will be mentally well and better able to fulfill those duties. Culturally competent mental health providers would be able to assess the cultural value of caballerismo—a positive male image that focuses on the man as the nurturing provider that defends the weak, is chivalrous and respects others—and utilize it to encourage fathers to seek treatment for their mental health issues so that they can continue to provide for their families instead of trying to stifle machismo beliefs. Therapists must be adept enough to demonstrate that they respect their client's cultural beliefs while getting them to consider other values that might be more supportive of good mental health.

Increasing Mental Health Knowledge

Most mental health providers, without hesitation, would suggest that increasing mental health knowledge is a clear way to reduce stigma. Research demonstrates that when people learn about the biological etiology of psychological disorders, this tends to result in reduced stigma for them [3]. Yet, this may be more complicated for

Latinx community members. Lopez and colleagues [43] found that among college students though greater depression knowledge was related to lower levels of stigma about mental health and reduced desire for social distance from someone with a disorder, it was also related to greater stigma about antidepressant medications. A similar finding was attained among a Latinx sample of depression clients in which their stigma for antidepressants increased with knowledge, purportedly as a result of learning about side effects [23]. Because of the strong cultural view that antidepressants are addictive, which may be increased with knowledge about side effects, this may be an area in which mental health providers must be very cautious when providing education. Antidepressants may be a last resort reserved only for those with severe depression and discussions of them must be handled carefully. However, in general, increasing knowledge about mental health can lead to reduced stigma with Latinx clients; therefore, psychoeducation should be a key aspect of all treatment.

Inclusion of Family in Treatment

Due to cultural values such as familismo and the interdependent nature of many in the Latinx community, support from family and friends may be a critical aspect of the treatment process for those with mental health stigma. Latinxs are more likely to disclose their mental health problems to family than friends [44], therefore the inclusion of family in the therapy process may be even more critical than including friends. Despite the likely cultural appropriateness of including family, this can be complicated due to stigma. A group of predominantly Puerto Ricans who utilized mental health treatment wanted their family members to participate in a psychoeducation program, but were fearful that they would not participate due to stigma [30]. But the inclusion of family in treatment or anti-stigma interventions programs could allow for conversations about mental illness which could lead to more positive family interactions. Psychoeducation or interventions may be an important first step before including family in the therapy process. Latinx clients will likely suffer additional distress if they feel they must hide their therapy from family members. It is critical that clinicians assist clients in navigating the process of disclosing their mental health treatment and inviting their family to join in their process.

Conclusion

There is much diversity within the American Latinx population that includes English language fluency, nationality, immigration status, acculturation, experiences with discrimination, and socioeconomic status among others. These factors must be considered by practitioners as they consider mental health stigma. There is no one size fits all approach. There is still much that we do not know about mental health stigma in the Latinx community, therefore researchers should strive to fill these gaps. Concerns that still need to be addressed more fully include a better understanding of heterogeneity among Latinxs, closer examination of the role of cultural values in understanding mental health stigma, issues within the measurement of mental health stigma, and examination of outcomes and concerns for disorders other than depression. Better knowledge about mental health stigma will assist mental health providers in offering more efficacious treatment to members of the Latinx community.

References

1. Dixon De Silva LE, Ponting C, Ramos G, Cornejo Guevara MV, Chavira DA. Urban Latinx parents' attitudes towards mental health: mental health literacy and service use. Child Youth Serv Rev. 2020 Feb 1;109:1–8.
2. Goodwill JR, Zhou S. Association between perceived public stigma and suicidal behaviors among college students of color in the U.S. J Affect Disord. 2020 Feb 1;262:1–7.
3. Corrigan PW, Druss BG, Perlick DA. The impact of mental illness stigma on seeking and participating in mental health care. Psychol Sci Public Interest. 2014;15(2):37.
4. Elkington KS, Hackler D, McKinnon K, Borges C, Wright ER, Wainberg ML. Perceived mental illness stigma among youth in psychiatric outpatient treatment. J Adolesc Res. 2012 Mar 1;27(2):290–317.
5. Cheng H, Kwan KK, Sevig T. Racial and ethnic minority college students' stigma associated with seeking psychological help: examining psychocultural correlates. J Couns Psychol. 2013;60(1):98–111. https://doi.org/10.1037/a0031169.
6. Whaley AL. Ethnic and racial differences in perceptions of dangerousness of person with mental illness. Psychol Serv. 1997;48(10):1328–130.
7. Chavira DA, Bantados B, Rapp A, Firpo-Perretti YM, Escovar E, Dixon L, et al. Parent-reported stigma and child anxiety: a mixed methods research study. Child Youth Serv Rev. 2017 May 1;76:237–42.
8. Turan JM, Elafros MA, Logie CH, Banik S, Turan B, Crockett KB, et al. Challenges and opportunities in examining and addressing intersectional stigma and health. BMC Med. 2019;17(7) https://doi.org/10.1186/s12916-018-1246-9.
9. Benuto LT, Gonzalez F, Newlands R, Casas J. ¡Yo no estoy loco/a! Improving treatment engagement for Latinos using an E-E video. J Health Disparities Res Pract. 2019 Jun 2;12(5):5–6.
10. DuPont-Reyes MJ, Villatoro AP, Phelan JC, Painter K, Link BG. Adolescent views of mental illness stigma: an intersectional lens. Am J Orthopsychiatry. 2020;90(2):201–11.
11. DeFreitas SC, Crone T, DeLeon M, Ajayi A. Perceived and personal mental health stigma in Latino and African American college students. Front Public Health. 2018 Feb;6 https://doi.org/10.3389/fpubh.2018.00049.
12. Turner EA, Jensen-Doss A, Heffer RW. Ethnicity as a moderator of how parents' attitudes and perceived stigma influence intentions to seek child mental health services. Cult Divers Ethn Minor Psychol. 2015 Oct;21(4):613–8.
13. Wang C, Barlis J, Do KA, Chen J, Alami S. Barriers to mental health help seeking at school for Asian–and Latinx–American adolescents. Sch Men Health: A Multidiscip Res Pract J. 2019 Sep 17; https://doi.org/10.1007/s12310-019-09344-y.
14. Collado A, Zvolensky M, Lejuez C, MacPherson L. Mental health stigma in depressed Latinos over the course of therapy: results from a randomized controlled trial. J Clin Psychol. 2019;75(7):1179–87.
15. Caplan S. A pilot study of a novel method of measuring stigma about depression developed for Latinos in the faith-based setting. Community Ment Health J. 2016 Aug;52(6):701–9.
16. Vega WA, Rodriguez MA, Ang A. Addressing stigma of depression in Latino primary care patients. Gen Hosp Psychiatry. 2010;32(2):182–91. Available from https://pubmed.ncbi.nlm.nih.gov/20302993/
17. Rojas-Vilches AP, Negy C, Reig-Ferrer A. Attitudes toward seeking therapy among Puerto Rican and Cuban American young adults and their parents. Int J Clin Health Psychol. 2011 May;11(2):313–41.

18. Mendoza H, Masuda A, Swartout KM. Mental health stigma and self-concealment as predictors of help-seeking attitudes among Latina/o college students in the United States. Int J Adv Couns. 2015;37(3):207.
19. Young AS, Rabiner D. Racial/ethnic differences in parent-reported barriers to accessing children's health services. Psychol Serv. 2015 Aug;12(3):267–73. https://doi.org/10.1037/a0038701.
20. Hirai M, Vernon LL, Popan JR, Clum GA. Acculturation and enculturation, stigma toward psychological disorders, and treatment preferences in a Mexican American sample: the role of education in reducing stigma. J Latina/o Psychol. 2015 May;3(2):88–102.
21. Higgins Neyland MK, Bardone-Cone AM. Treatment experiences of Latinas with current or past binge eating disorder and/or bulimia nervosa. Eat Disord. 2019 May;27(2):253–65.
22. Clement S, Schauman O, Graham T, Maggioni F, Evans-Lacko S, Bezborodovs N, et al. What is the impact of mental health-related stigma on help-seeking? A systematic review of quantitative and qualitative studies. Psychol Med. 2015;45(1):11–27.
23. Sanchez K, Killian MO, Eghaneyan BH, Cabassa LJ, Trivedi MH. Culturally adapted depression education and engagement in treatment among Hispanics in primary care: outcomes from a pilot feasibility study. BMC Fam Pract. 2019 Oct 21;20(1) https://doi.org/10.1186/s12875-019-1031-7.
24. Diala C, Muntaner C, Walrath C, Nickerson KJ, LaVeist TA, Leaf PJ. Racial differences in attitudes toward professional mental health care and in the use of services. Am J Orthopsychiatry. 2000;70(4):455–64.
25. Quinn DM, Camacho G, Pan-Weisz B, Williams MK. Visible and concealable stigmatized identities and mental health: experiences of racial discrimination and anticipated stigma. Stigma and Health. 2019 Dec 12;5(4):488–91. https://doi.org/10.1037/sah0000210.
26. Interian A, Ang A, Gara MA, Link BG, Rodriguez MA, Vega WA. Stigma and depression treatment utilization among Latinos: utility of four stigma measures. Psychiatr Serv. 2010 Apr;61(4):373–9.
27. Vargas SM, Cabassa LJ, Nicasio A, De La Cruz AA, Jackson E, Rosario M, Guarnaccia PJ, Lewis-Fernandez R. Toward a cultural adaptation of pharmacotherapy: Latino views of depression and antidepressant therapy. Transcult Psychiatry. 2015;52(2):244–73.
28. Caplan S, Paris M, Whittemore R, Desai M, Dixon J, Alvidrez J, et al. Correlates of religious, supernatural and psychosocial causal beliefs about depression among Latino immigrants in primary care. Ment Health Relig Cult. 2010;13:1469–74. https://doi.org/10.1080/13674676.2010.497810.
29. Choi N-Y, Kim HY, Gruber E. Mexican American women college students' willingness to seek counseling: the role of religious cultural values, etiology beliefs, and stigma. J Couns Psychol. 2019 Oct;66(5):577–87.
30. Hackethal V, Spiegel S, Lewis-Fernández R, Kealey E, Salerno A, Finnerty M. Towards a cultural adaptation of family psychoeducation: findings from three Latino focus groups. Community Ment Health J. 2013 Oct;49(5):587–98.
31. DuPont-Reyes MJ, Villatoro AP, Phelan JC, Painter K, Link BG. Media language preferences and mental illness stigma among Latinx adolescents. Social psychiatry and psychiatric epidemiology. Int J Res Soc Genetic Epidemiol Ment Health Ser. 2019, 1
32. Villatoro AP, Morales ES, Mays VM. Family culture in mental health help-seeking and utilization in a nationally representative sample of Latinos in the United States: the NLAAS. Am J Orthopsychiatry. 2014 Jul;84(4):353–63.
33. Abdullah T, Brown TL. Mental illness stigma and ethnocultural beliefs, values, and norms: an integrative review. Clin Psychol Rev. 2011 Aug;31(6):934–48.
34. Corrigan PW, Morris SB, Michaels PJ, Rafacz JD, Rüsch N. Challenging the public stigma of mental illness: a meta-analysis of outcome studies. Psychiatr Serv. 2012;63:963–73. https://doi.org/10.1176/appi.ps.201100529.
35. Painter K, Phelan JC, Dupont-Reyes MJ, Barkin KF, Villatoro AP, Link BG. Evaluation of Antistigma interventions with sixth-grade students: a school-based field experiment. Psychiatr Serv. 2017 Apr;68(4):345.

36. Cabassa LJ, Molina GB, Baron M. Depression Fotonovela: development of a depression literacy tool for Latinos with limited English proficiency. Health Promot Pract. 2012 Nov;13(6):747–54.
37. Cabassa LJ, Oh H, Humensky JL, Unger JB, Molina GB, Baron M. Comparing the impact on latinos of a depression brochure and an entertainment-education depression "Fotonovela.". Psychiatr Serv. 2015;3:313.
38. Sánchez M, Cardemil E, Adams ST, Calista JL, Connell J, DePalo A, et al. Brave new world: mental health experiences of Puerto Ricans, immigrant Latinos, and Brazilians in Massachusetts. Cult Divers Ethn Minor Psychol. 2014 Jan;20(1):16–26.
39. Hernandez MY, Organista KC. Entertainment-education? A fotonovela? A new strategy to improve depression literacy and help-seeking behaviors in at-risk immigrant Latinas. Am J Community Psychol. 2013 Dec;52(3–4):224–35.
40. Unger JB, Cabassa LJ, Molina GB, Contreras S, Baron M. Evaluation of a Fotonovela to increase depression knowledge and reduce stigma among Hispanic adults. J Immigr Minor Health. 2013;15(2):398–406.
41. Lin L, Stamm K, Christidis P. How diverse is the psychology workforce? Monit Psychol. 2018;2. https://www.apa.org/monitor/2018/02/datapoint
42. American Psychological Association. Ethical principles of psychologists and code of conduct. 2017. https://www.apa.org/ethics/code/
43. Lopez V, Sanchez K, Killian MO, Eghaneyan BH. Depression screening and education: an examination of mental health literacy and stigma in a sample of Hispanic women. BMC Public Health. 2018 May 22;18(1):646.
44. Chang J, Chen C-N, Alegría M. Contextualizing social support: pathways to help seeking in Latinos, Asian Americans, and whites. J Soc Clin Psychol. 2014 Jan;33(1):1–24.

Psychotic Disorders in Hispanics

4

Tatiana Falcone, Ruby Castilla-Puentes, and Alireza Minagar

Introduction

Psychotic disorders constitute a wide gamut of medical diseases (e.g., psychiatric, neurologic), which share certain clinical manifestations such as hallucination, delusions, and significant decline in the patient's intellectual and professional abilities and achievements. These illnesses present with different epidemiology, and underlying pathophysiology, and their diagnosis and treatment depend on the underlying cause. As of 2019, approximately 55 million Hispanics live in the USA, which roughly constitutes 18.5% of the population [1]. The three largest Latino groups in the USA include Mexicans, Cubans, and Dominicans [2]. Psychotic disorders have been partially studied across different racial and ethnic groups with efforts to define the variability of frequency of occurrence of certain clinical manifestations among these patients. With the growing trend of the Hispanic population in the USA, clinicians should learn about the presentations and therapy of this subpopulation. The present chapter discusses psychotic disorders and explores their occurrence and unique clinical features focus in Hispanic American patients. It is also noteworthy that, in this chapter, the terms "Hispanic," "Latino," and "Hispanic American" are used interchangeably to describe the Hispanic population.

T. Falcone
Department of Psychiatry, Cleveland Clinic, Cleveland, OH, USA

R. Castilla-Puentes (✉)
Janssen Research & Development LLC, Titusville, NJ, USA

Center for Clinical and Translational Science and Training, University of Cincinnati, Cincinnati, OH, USA

WARMI Mental Health, Collaborative Mental Health Network, Cincinnati, OH, USA
e-mail: RCastil4@its.jnj.com

A. Minagar
Department Bioinformatics, University of Maryland Global Campus, Largo, MD, USA

R. Castilla-Puentes, T. Falcone (eds.), *Mental Health for Hispanic Communities*,
https://doi.org/10.1007/978-3-031-13195-0_4

Schizophrenia

Schizophrenia, as a heterogeneous clinical syndrome, constitutes one of the most serious psychiatric illnesses. The assemblage of manifestations of schizophrenia is classified in three areas of positive, negative symptoms and cognitive impairment. Diagnosis of schizophrenia rests on a comprehensive interview and assessment of the patient and utilization of DSM-5 diagnostic criteria. Schizophrenia occurs globally with a prevalence of approximately 1% and an incidence of 1.5 per 10,000 individuals and is usually diagnosed in the late teen years to the early part of the third decade of life [3]. It tends to present earlier in males (modal age 18–25) than females (modal age 25–35 with a second peak around the menopause) and interestingly subtle alterations of cognition and impairment of social functions and relationship may manifest, even year, prior to psychiatric diagnosis is made. Almost half of the schizophrenia patients suffer from concurrent mental and/or behavioral abnormalities [4]. Hereditability and family history of schizophrenia is a significant risk factor, while the role of other factors such as season and birthplace, socioeconomic status, and maternal infections remain hypothetical. Epidemiologic assessments of patients with schizophrenia and bipolar disorder with psychotic features demonstrate 50% concordance of certain genetic loci between identical twins as well as among parents and siblings [5].

Pathophysiology of schizophrenia includes dysregulation and potential dysfunction of neuroanatomic pathways involving dopaminergic system, GABAergic, and hypofunction of the N-methyl-d-aspartate receptor [6]. In addition, functional abnormalities of the dorsolateral prefrontal cortex circuit have been implicated in the development of schizophrenia. One recent study revealed that enhanced expression of schizophrenia-associated gene C4 is associated with hypoconnectivity of the prefrontal cortex and decreased social interaction [7].

Schizophrenic patients commonly reveal affective symptoms as decreased emotional responsiveness (sometimes called anhedonia) or overly active and improper emotions such as excessive rage, delightedness, and anxiety. Patients with flat or blunted affect may suffer from antipsychotic-associated parkinsonism, profound depression, or it may be due to underlying illness. Additionally, patients report feelings of omnipotence, perplexity, a feeling of seclusion, and being ambivalent.

Catatonic patients demonstrate muteness, negativism, and automatic conformity. Waxy flexibility, as a form of mannerism, is uncommonly seen in patients with schizophrenia.

As a disease, schizophrenia involves and distorts the process or the content of the patient's thoughts and the clinician can detect thought disorder during the psychiatric interview. Patients with schizophrenia and thought abnormality may show tangential thinking, neologisms, echolalia, word salad, mutism, and circumstantiality. Delusions, a cardinal feature of schizophrenia and as a sign of disorder of thought content, vary and may present as delusions of control, grandiosity, guilt, reference, persecution, thought insertion, withdrawal, or broadcasting, and somatic form. Another frequently encountered thought problem is hallucinations. Schizophrenic patients commonly suffer from hallucinations and the most common one is auditory, when the patient may hear threatening, obscene, critical, or offensive voices.

Some patients report two dissimilar voices conversing among themselves or a single voice commenting on the patient's behavior [8]. Visual hallucination is the next common form. Tactile, olfactory, or gustatory ones are less frequent, and once present, the possibilities of an underlying medical or neurologic illness should be entertained [9].

Patients with schizophrenia are usually fully oriented during the interview; however, many suffer from cognitive decline. While memory remains relatively intact, further probing reveals subtle cognitive impairment with a decline in domains of working memory, executive function, attention, and episodic memory. In general, cognitive decline in schizophrenia is not progressive.

In most cases, schizophrenic patients demonstrate poor insight into the nature and depth of their illness, which is further complicated by poor compliance with treatment. The treating physician should particularly assess the patient for his or her understanding of the diagnosis and reliability and compliance with treatment methods. Schizophrenia patients should be assessed for safety concerns regarding suicide and homicide.

The criteria of DSM-5 are used for the diagnosis of schizophrenia. Briefly, diagnosis of schizophrenia requires that the patient shows any two or more of the psychotic symptoms (delusions, hallucinations, disorganized speech, grossly disorganized or catatonic behavior, negative symptoms) for at least 6 months (except if the patient has been treated). Each of these symptoms which include delusions, hallucinations, and disorganized speech should be present for a 1-month period (for ICD-10) and 6 months (for DSM-5). The presence of delusion or hallucination is not necessarily required for a diagnosis of schizophrenia. In addition, in all patients suspected of having schizophrenia, schizoaffective disorder and bipolar disorder with psychotic features should be excluded. Substance use disorder, use of certain medications which cause psychosis, and underlying medical diseases should be excluded. When evaluating patients with catatonia, clinicians should look carefully for a past history of manic or depressive episodes and for a family history of mood disorders. Manic symptoms in persons from minority groups (particularly blacks and Hispanics) are often misdiagnosed as schizophrenic symptoms [10].

Neuropsychological assessment is not required for the formation of a diagnosis, though it can reveal the cognitive decline and assist in targeted cognitive treatment.

Differential Diagnosis of Schizophrenia

A wide spectrum of medical and psychiatric conditions come in the differential diagnosis of schizophrenia. These include the central nervous system (CNS) tumors and trauma, ischemic stroke, CNS metal toxicity, epilepsy, viral encephalitis, AIDS, neurosyphilis, Huntington's disease, Lyme disease, acute intermittent porphyria, metachromatic leukodystrophy, and neuropsychiatric syndrome of systemic lupus erythematosus. Major psychiatric disorders coming to the differential diagnosis of schizophrenia consist of autism spectrum disorder, brief psychotic disorder, schizophreniform disorder, mood disorders with psychotic features, schizoaffective disorder, drug-associated psychosis, and delusional disorder.

Schizophreniform Disorder

Schizophreniform disorder recognized as a psychotic disorder while the psychotic symptoms (delusions and hallucinations), with or without the presence of the negative symptoms, last more than one month and less than 6 months and this is followed by recovery.

Schizoaffective Disorder

Schizoaffective disorder, with a lifetime prevalence of less than 1%, is characterized by the development of prominent mood symptoms such as depression or mania during the course of chronic psychotic disorder. Schizoaffective disorder probably is a heterogeneous disorder with multiple underlying etiologies. The course and outcome of the schizoaffective disorder varies depending on whether it clinically presents with episodes of bipolar cycling or the presence of bouts of depression without manic episodes. Women are more frequently affected than men.

Diagnosis of schizoaffective disorder needs that a continuous period of illness (at least 6 months) with major mood episodes (manic or depressive attack) should happen simultaneously with core psychotic symptoms (criterion A of schizophrenia) and last for most of the entire course of psychotic periods. When the abnormal mood is absent, then delusions or hallucinations should be present for at least two weeks. The clinical symptoms of major mood disorder are present for most of the entire length of the active and residual segments of the illness. No better diagnoses or explanation for the clinical disorder can be identified. Patients with schizoaffective disorder possess a 5% risk of suicide and the existence of depression enhances this risk. In addition, this disorder can significantly impair the social, occupational, and academic functions of the patient.

Both schizophreniform disorder and schizoaffective disorders are among the differential diagnoses of schizophrenia.

Mood Disorders

Patients with bipolar disorder and those with major depressive episodes may manifest delusions and hallucinations. Patients suffering from psychotic depression demonstrate delusions that are consistent with their depressed mood such as guilt, warranted punishment, and self-criticizing. Patients with mood disorders with psychotic features do reveal a decline in self-care, isolation, and loss of ability to function in society. The opposite scenario also exists and manic patients frequently reveal delusions and hallucinations. Interestingly, delusions in manic patients are congruent with their underlying mood disorder and suffer from grandiosity and flight of ideas. Detailed psychiatric assessment assists the clinician to differentiate these mood disorders from schizophrenia.

Psychosis as a Manifestation of Medical Illness

It is well recognized that patients with medical conditions manifest with psychiatric symptoms such as psychosis, agitation, alterations of personality, mania, and cognitive decline. Patients with neurologic diseases, infectious illnesses, endocrine diseases, malignancies, and metabolic conditions present with psychosis. The encephalitis due to anti-NMDA receptor autoantibodies, which is an example of an autoimmune neurologic disorder with psychiatric presentations present with delusions, hallucination, or mania.

Neurologic Diseases

Patients with stroke, multiple sclerosis, and epilepsy can show psychiatric manifestations. Both ischemic and hemorrhagic strokes present with a wide range of psychiatric symptoms including psychosis, cognitive impairment, and mood changes.

Both ischemic and hemorrhagic strokes happen with greater frequency in Hispanics than non-Hispanics. Ischemic strokes produce a range of neuropsychiatric manifestations and disorders including delusional disorder, schizophrenic-like psychosis, mood disorder with psychotic features and hallucinations. Stroke lesions which involve certain neuroanatomic regions such as the right hemisphere (particularly frontal, temporal, and parietal areas) and the right caudate nucleus produce neuropsychiatric syndromes and their presence may be associated with a less favorable prognosis. Patients with post-stroke psychosis may or may not respond to therapy with antipsychotics (Stangeland JNNP).

Patients suffering of epilepsy may present with psychosis, particularly during post-ictal and interictal periods. Patients with complex partial seizures, particularly cases where the seizures involve the temporal lobes, present with psychotic episodes, which may resemble schizophrenia. Certain features such as the early onset of epileptic fits, lesions located in the medial temporal lobe, and the presence of seizure on the left side are accompanied by the manifestation of psychosis [11]. Uncommonly, schizophrenia patients may have concurrent complex partial epilepsy. Antiepileptics may be associated with psychiatric features.

Multiple sclerosis (MS), an immune-mediated demyelinating and neurodegenerative disease of the human central nervous system, presents with a spectrum of neuropsychiatric syndromes, which include cognitive impairment and memory loss, dementia, depression, mania, delusion, and hallucination. In addition, some of the medications such as corticosteroids and beta-interferons which are used for the treatment of MS patients possess their own psychiatric side effects including manic behavior and depression, respectively [12].

Infectious Illnesses

Both systemic and central nervous system infections present with psychiatric presentations. Patients with untreated syphilis may present with psychiatric manifestations such as cognitive decline, personality disorders, delirium confusion, paranoia, hallucinations, mood alterations including mania ([13], J neuropsychiatry Clin Neuroscin).

Patient with AIDS demonstrates various psychiatric symptoms and syndrome which stem from the direct impact underlying HIV infection or other variables such as pre-existing psychiatric disorders, substance use disorder, opportunistic infection, or even adverse effects of medical therapies. Major behavioral and psychiatric disorders among these patients include mania, depression, anxiety, delusion, hallucination, and apathy.

Patients with herpes simplex virus present with agitation, bizarre behavior, and psychosis. Herpetic encephalitis affects the temporoparietal regions of the human brain and patients develop focal or generalized seizures, olfactory hallucinations, personality changes, psychosis, and dysphasia or aphasia. Rarely, patients with herpes simplex encephalitis develop Kluver-Bucy syndrome with bilateral involvement of the temporal lobes and demonstrate hypersexuality, hyperorality, irritability, and psychosis.

Prion diseases, particularly Creutzfeldt–Jakob disease, are uncommon lethal neurodegenerative pathologies of the human brain and clinically present with rapidly progressive dementia, alterations of personality, obsessive-compulsive symptoms, hallucinations, and psychosis. Variant CJD with a slower course of deterioration presents with CJD may be familial, sporadic, or acquired [14].

Systemic lupus erythematosus (SLE), an immune-mediated vasculitis, presents with neuropsychiatric disorders which consist of cognitive impairment, anxiety and mania, and depression. Patients may also develop corticosteroid-induced psychosis.

Endocrine and Metabolic Diseases

Vitamin B1 and B12 deficiencies manifest with a wide gamut of psychiatric manifestations such as dementia, psychosis, mania, and mood alterations. Vitamin B1 (thiamine) deficiency, common among alcoholics and patients with malnutrition or malabsorption, presents with a Wernicke-Korsakoff syndrome, which clinically manifests with ataxia, nystagmus, intranuclear ophthalmoplegia, psychosis, confusion, and agitation.

Malignancies

Patients with the central nervous system (both primary and metastatic) and systemic cancers present with psychiatric symptoms including psychosis, mania, depression, cognitive decline and dementia, mood disorders, and personality changes.

Paraneoplastic syndromes indicate how the neoplasms outside the central nervous system can affect its activities without direct invasion or metastasizing. Paraneoplastic syndrome is frequently associated with lung, breast, and ovarian malignancies. A common syndrome is limbic encephalitis which presents with short-term amnesia, sleep disturbances, delusions, hallucinations, seizures, and psychosis.

Drug-Induced Psychosis

Both use and withdrawal from certain medications and illegal substances may be associated with psychosis. Some of these medications include sedative-hypnotics, alcohol, and anxiolytics.

Patients withdrawing from alcohol, benzodiazepines, and barbiturates may manifest psychosis, delirium, seizures, hypertension, and agitation.

In patients with drug-induced psychosis, the treating physician should seek for signs and symptoms of acute intoxication or chronic utilization and the psychotic features should resolute with abstinence. Part of the assessment of psychiatric patients should include blood and urine toxicology screening to exclude drug use and drug-induced psychosis. The list of medications and substances which are associated with the generation of psychosis is extensive and the most significant ones consist of antidepressants, anticholinergics, dopamine agonists (such as pramipexole and L-dopa), hallucinogens (such as phencyclidine, cannabis, mescaline, and LSD), histamine-2 antagonists, psychostimulants (such as cocaine and amphetamine), and sympathomimetics.

Cannabis, as the most frequently used illegal substance in many countries, generates temporary and usually mild psychotic and affective experiences in healthy users. Once an individual smokes marijuana, euphoria develops within minutes, peaks in almost 30 minutes and such effect lasts 2 to 4 hours. Cannabis use and cannabis intoxication are associated with psychosis (cannabis-induced psychosis), which come in the differential diagnosis of schizophrenia [15]. In addition, many schizophrenias use cannabis. There is not much data about the use of cannabis across various ethnic groups. Based on the existing data on Hispanic Americans, in 2018, 8.6 million Hispanic adults had a mental and/or substance use disorder. According to American Addiction Centers, 7.1% of Hispanic Americans suffer from a substance use disorder, compared to 7.4% rate of substance use disorder in the general population. Approximately, one-third of Americans meet the diagnostic criteria for alcohol use disorder on a lifetime basis [16].

Amphetamines such as methamphetamine and 3,4 methylenedioxymethamphetamine (MDMA) belong to a group of chemical compounds known as phenethylamine and exert catecholaminergic effects on the CNS and peripheral circulation. While amphetamines are clinically used for treatment of certain medical conditions such as narcolepsy and attention-deficit disorder with hyperactivity, they are also used recreationally. The use of amphetamines can be associated with acute psychosis with visual hallucination, persecutory delusions, and the clinical presentation simulates schizophrenia [17].

Certain medications such as anabolic-androgenic steroids, estrogens, and corticosteroids can induce euphoria, hyperactive behavior, alterations of mood, violence, and hallucinations. Patients under the effect of these medications may develop psychosis and even attempt suicide [18].

Inhalants (such as gasoline, toluene, and spray paint) are a group of substances that can cause delirium and disturbing behavior. They can also cause psychosis and the patient shows paranoid behavior, anxiety, and agitation [19].

Brief Psychotic Disorder

As a disorder on the list of differential diagnoses of schizophrenia, the brief psychotic disorder usually presents with an acute demonstration of psychotic symptoms, particularly following a social or psychological stressor associated with fair premorbid functioning. Recently, a brief psychotic disorder associated with COVID-19 infection has been reported [20]. It is expected that brief psychotic disorder resolves swiftly and totally in response to treatment with antipsychotics. The social stressors and precipitating factors should be identified and eliminated. In most cases, inpatient treatment is indicated followed by psychopharmacologic therapy. Antipsychotics and sedatives are utilized to relieve the acute symptoms and improve the patient's rest and sleep.

Treatment of Psychotic Disorders

Treatment of patients consists of three main objectives: symptomatic treatment, maintenance and promotion of general health and sense of well-being, and reducing the debilitating psychosocial impairments. Antipsychotic medications are utilized to manage the clinical features and circumvent relapses. These medications, in many cases, effectively control disease manifestations and decrease relapses. Antipsychotics are categorized as first generation versus the second generation. Chlorpromazine and fluphenazine represent the first generation while aripiprazole and asenapine are examples of the second generation of antipsychotics. Tables 4.1 and 4.2. contain the list, dosage, and side effects, and other issues regarding utilizing these medications. Both first and second generation of antipsychotic medications possess unique side effects and complications. Briefly, neurologic adverse effects of antipsychotics consist of acute dystonia, akathisia, parkinsonian syndrome, neuroleptic malignant syndrome, tardive dyskinesia, and peri-oral tremor. These issues are further expanded in Table 4.1.

Indications for hospitalizations include acute exacerbations (e.g. when psychotics symptoms are severe and cause distress), the presence of disturbing behaviors, the risk of suicide, and when the patient demonstrates aggressive violent behavior.

Uncommonly, electroconvulsive therapy (ECT) is employed for the treatment of schizophrenic patients with severe unrelenting symptoms of catatonia, and insufficient response to pharmacotherapy.

Psychological evaluation and rehabilitative therapy include cognitive and behavioral therapy, psychoeducation, cognitive remediation, and providing supportive and motivational interventions for the patients and their families. Once the acute psychosis is stable and the patient is no longer violent and does not present to self or others, psychological therapy with emphasis on social skills training initiates. Psychological training and education of the family member is another approach to empower family members in support of the patient. Cognitive rehabilitation may be

Table 4.1 List of FDA-approved first-generation antipsychotics

Medication name First-generation antipsychotic	Dosage	Comments
Chlorpromazine	300–1000 mg	In all patients before initiating treatment with antipsychotics, treating physician should establish baseline ECG, serum glucose and lipid profile and then at month 3 and then annually unless otherwise needed
Droperidol		
Fluphenazine	5–20 mg	
Haloperidol	5–20 mg	
Loxapine	60–100 mg	
Perphenazine	16–64 mg	
Pimozide	2–4 mg	
Thiothixene	20–30 mg	
Trifluoperazine	15–50 mg	
Thioridazine	200–800 mg	

Note: Antipsychotics are approved by the FDA for a number of indications which consist of schizophrenia, bipolar affective disorder, major depressive disorder, agitation accompanied with schizophrenia and bipolar mania, autism, Tourette's syndrome, and psychosis
Doses cited are the maintenance dosages

Table 4.2. List of FDA-approved second-generation antipsychotics

Medication name Second-generation antipsychotic	Dosage	Comments
Aripiprazole	10–30 mg	Start the antipsychotic at lowest dose Be on alert for extrapyramidal syndrome (more common with first-generation antipsychotics) Endocrine adverse effects including hyperprolactinemia Weight gain, abnormalities of serum glucose and lipid profile Once the patient on antipsychotic for 2 weeks or more, assess the efficacy
Asenapine	5–10 mg	
Clozapine	250–450 mg	
Iloperidone	6–12 mg	
Olanzapine	10–20 mg	
Quetiapine	400–800 mg	
Paliperidone	3–12 mg	
Risperidone	2–6 mg	
Ziprasidone	80–160 mg	

beneficial for the resolution of psychotic features and help psychotic patients, such as those with schizophrenia, to restructure their beliefs and decrease stress due to hallucinations and delusions [21].

Usually, patients benefit from taking healthy diets and regular exercise. Patients with schizophrenia tend to be more obese and carry a higher risk of diabetes mellitus type-II compared to the general population. Elevated risks of diabetes mellitus, obesity, smoking, and other substance use disorders increase the risk of cardiovascular morbidity and mortality.

Prognosis of schizophrenia is complicated and assists treating physicians in the treatment process. Patients with schizophrenia present with an elevated risk of substance abuse, which in turn is associated with more often inpatient admissions, first hospitalization at a younger age, and more compromised individual and family impairment. Firmer commitment to the delusions and more compromised executive function increases the chance of rehospitalization and more need for the outpatient clinics [22]. Severe psychotic features are associated with more aggression and divorce is more common in schizophrenia. Patients with schizophrenia have a higher rate of accelerated cardiac disease, which is partly due to an elevated risk of cigarette smoking in this population. Schizophrenia patients have a 10% lifetime risk of suicide and certain issues such as drug abuse, profound depression, earlier suicide attempts, poor compliance with treatment, and restless behavior elevate the chance of suicide attempt [23]. As comorbid conditions, substance use disorders have lifetime prevalence over 70% in patients with schizophrenia. Many patients develop dependence on tobacco, alcohol, cannabis and cocaine. Certainly, co-existing substance use disorder(s) complicates the therapy and does not improve the prognosis.

Psychotic Disorders Among Hispanic Americans

Hispanic Americans is perhaps the largest and most rapidly growing minority groups in the USA and it is expected that, by 2060, this ethnic group makes up to 28.6% of US population [1]. Among the Hispanic Americans, there are diverse subgroups such as Mexican Americans, Cuban Americans, and Puerto Ricans. For decades, a number of epidemiological studies have assessed the rates of psychiatric disorders among various ethnic groups, particularly Hispanic Americans and African Americans [24]. These studies have also attempted to identify the diagnosis rate of these disorders and access to the healthcare of Hispanic Americans.

It is believed that cultural factors influence or even form the pattern, content, and extent of clinical manifestations of psychiatric disorders. Assessment of the Hispanic Americans reveal the existence of certain ethnic and racial prejudiced opinions regarding psychiatric disorders. For example, many Latinos consider mental and psychiatric disorders as sign of weakness and some even use disparaging terms such as "loco" when they refer to these patients. The underlying culture also may impact the presentation of psychiatric disorders. The study by Weisman (Weisman, Lopez, Ventura) compared the psychiatric symptoms between White Americans and Hispanic Americans of Mexican origin. The authors reported that Hispanic Americans of Mexican origin reported more often physical symptoms than White Americans. They reported a lower frequency of psychiatric symptoms such as persecutory delusions, blunted affect, and nervous tension. Assessing the psychotic symptoms in Hispanic Americans is difficult, particularly in patients with mood and anxiety disorders because their manifestation may differ from those encountered in non-Hispanic patients. Hispanic patients with anxiety and depression report atypical auditory and visual hallucinations, while the mental status is otherwise unremarkable with no evidence of thought disorder. Such effect of culture in the clinical presentation of psychotic disorders in Hispanic Americans may make the correct diagnosis of psychotic disorder a diagnostic challenge [25]. Anglin and Malaspina [26] evaluated the ethnicity effects on clinical diagnoses compared to best-estimate diagnoses in patients with psychosis. During this retrospective medical chart review, the authors compared the two methods. The findings of the study demonstrated in Hispanic patients than African American patients and also found that diagnostic concurrence for paranoid schizophrenia was the best in Hispanic patients.

Schizophrenia carries the same prevalence rate between Hispanics and non-Hispanics; however, dissimilarities in clinical presentations have been observed in certain subgroups of Hispanic patients compared to non-Hispanics.

Psychotic symptoms occur in the context of post-traumatic stress disorder (PTSD) and they have been reported to occur more frequently in Hispanic veterans [27]. Hispanic Americans compared to Whites are less likely to receive minimal trials of psychopharmacological treatments for PTSD. Such a shortcoming in treatment of these patients further complicates the process of learning more about the epidemiologic features of PTSD among Hispanics in the United States.

An interesting study by Escamilla et al. [28] concentrated on defining evidence in favor of association of the Epsin 4 gene to schizophrenia in a cohort of 1423 subjects from 337 Latino families who were genotyped utilizing third-degree single nucleotide polymorphisms (SNPs) spanning Epsin 4 gene. The investigators reported variation in Epsin 4 gene showed a significant association with psychotic disorder in the Latino population.

Substance use disorders are common among various ethnic groups in the USA and the use of many of these compounds is associated with delusion, hallucination, and psychosis. In addition, some of these substances such as marijuana are considered "gateway drugs" and their use is associated with an elevated risk of progression to other illegal drugs. Yet, race and ethnicity also play role in the rate of their use. Assessment of the rate of marijuana use among White Americans, African Americans, and Hispanic Americans reveals that in the age group 17 to 34, Hispanics' use of marijuana is less than the other two ethnic groups. Again, in the age group 35 years and older, White Americans and African Americans show higher rates than Hispanic Americans. Hallucinogen drugs are used by various racial and ethnic groups. White Americans use hallucinogens more than African Americans (2:1), while the ratio for White to Hispanics is 1.5:1.

Another epidemiologic study (based on the findings from the 2015–2018 National Survey of Drug Use and Health) assessed cocaine use among Hispanics [29]. The investigators utilized weighted analyses to identify correlates to past-year cocaine use and their findings showed that 4.11% of Hispanic young adults used cocaine during the past year. A separate epidemiologic study of cocaine and overdose mortality revealed a less rate of cocaine-associated overdose mortality in Hispanic Americans compared to African Americans [30].

A significant subject is the impact of migration and citizenship status on Hispanics who live in America. Migration from one country to another is inherently a stressful event and may affect the individual's mental status and may act as a predisposing factor to the emergence of psychiatric disorders in Hispanics with uncertain immigration status.

Socioeconomically, Hispanic American minority have more restricted revenue, with fewer access to educational opportunities, and more difficult access to the healthcare system. Such social disparities, in turn, may promote noncompliance and less reliable follow-up care. One should bear in mind that many Hispanic Americans with low income have no or minimal health coverage, which in turn limits their access to mental health clinics. Other factors which restrict these patients' access to mental health include the absence of knowledge regarding available mental and health services, language difficulties which bars them from properly expressing their symptoms and seeking appropriate medical advice and treatment, concerns, and anxiety about local and federal agencies and law enforcement, and significant discomfort of being diagnosed with psychiatric disorders and feel they are marked with certain social stigmata.

Even treatment of psychotic disorders may be influenced by ethnic parameters. Puyat et al. [31] presented a systematic review and meta-analysis assessing the racial and ethnic disparities in the utilization of antipsychotics. The authors reported no significant disparities and inequalities in the utilization versus non-use of antipsychotics; however, they noted that ethnic minorities (including Hispanic Americans) were continuously less treated with newer antipsychotics than non-ethnic groups. A significant issue is the racial disparities in the diagnosis of psychiatric disorders. The authors reported a clear and pervasive pattern of Latino American/Hispanic patients was disproportionately diagnosed with a psychotic disorder on average three times more commonly than Euro-American/White patients. Possible explanations for such disparity in clinical diagnosis may stem from clinical bias and sociological roots such as unequal access to the healthcare service and tendency to participate in the mental healthcare system. In another study, Oluwoye et al. [32] assessed the role of racial and ethnic differences in therapy outcomes among participants via a randomized controlled trial of a therapeutic intervention for the first-episode psychosis. The investigators reviewed the impact of race and ethnicity on psychiatric symptoms and psychological education. According to the findings of this clinical study, Hispanic subjects were more likely than non-Hispanic White participants to receive medication management. A separate study reviewed the family processes and therapy seeking among Latinos with first-episode psychosis [33]. The authors assessed family processes before the treatment and their relation to the duration of untreated psychosis among Latino individuals with the first-episode psychosis. The results of the study indicated the relevance and significance of family context in the provision of early therapy for Latino patients with first-episode psychosis. Based on their findings, family-based supportive services were required to enhance knowledge of the manifestation of psychosis and afford families with the support that enables patient therapy during the critical period of psychosis. Another clinical observation by Chartier et al. [34] entailing three Hispanic subgroups (Cuban American, Mexican American, and Puerto Ricans) explored their dissimilarities in substance use therapy and the relationship of acculturation features to the therapy outcomes. Based on the findings of this multicenter randomized clinical trial of motivational therapy versus usual therapy, the investigators reported that Cuban Americans and those study participants who were more connected to Hispanic culture demonstrated lower therapy retention. Hispanic patients who were born in the USA and those who spoke English revealed a less percentage of the number of days abstinent during weeks 5–16. However, Puerto Ricans and Cuban Americans who has stayed in the USA for an extended period demonstrated higher rates of days being sober in weeks 1–4 and 5–16, respectively. The study emphasized the role of the connection between acculturation and therapy outcome.

Conclusion

Psychotic disorders, while present as interesting illnesses with unrelated underlying mechanisms include a vast number of disorders. The common feature among this is the fact that the patient disconnects from the world of reality and manifests hallucination, delusion, and abnormalities of thought content which often compromise patient's academic achievement, ability to work, and capability to interact with other members of the society. Some of the psychotic patients particularly during the acute phase become violent, may pose threat to self or others, and may take their life. Both ethnic and cultural factors play roles in clinical manifestations of psychotic disorders. In addition, epidemiologic studies have revealed racial and ethnics differences in the epidemiology and manifestations of these disorders. Psychotic disorders in Hispanic Americans, as a rising ethnic group in the USA, are significantly underexplored and certain factors such as language barrier and personal understanding of these disorders as well as socioeconomic factors may restrict this subpopulation access to mental healthcare. More research effort, funding, and more resources, including bioinformatics, pharmacogenomic, artificial intelligence, and biotechnologic tools, are required to appropriately perform culturally based studies into prevalence, clinical presentations, and treatments for psychotic disorders among Hispanic population in the USA.

References

1. Geltman D, Chang G. Hallucinations in Latino psychiatric outpatients: a preliminary investigation. Gen Hosp Psychiatry. 2004;26(2):153–7.
2. Aguayo-Mazzucato C, Diaque P, Hernandez S, Rosas S, Kostic A, Caballero AE. Understanding the growing epidemic of type 2 diabetes in the Hispanic population living in the United States. Diabetes Metab Res Rev. 2019;35(2):e3097.
3. McGrath J, Saha S, Chant D, Welham J. Schizophrenia: a concise overview of incidence, prevalence, and mortality. Epidemiol Rev. 2008;30:67–76.
4. Tsai J, Rosenheck RA. Psychiatric comorbidity among adults with schizophrenia: a latent class analysis. Psychiatry Res. 2013;210(1):16–20.
5. Gejman PV, Sanders AR, Duan J. The role of genetics in the etiology of schizophrenia. Psychiatr Clin North Am. 2010;33(1):35–66.
6. Balu DT. The NMDA receptor and schizophrenia: from pathophysiology to treatment. Adv Pharmacol. 2016;76:351–82.
7. Comer AL, Jinadasa T, Sriram B, Phadke RA, Kretsge LN, Nguyen TPH, Antognetti G, Gilbert JP, Lee J, Newmark ER, Hausmann FS, Rosenthal S, Liu Kot K, Liu Y, Yen WW, Dejanovic B, Cruz-Martín A. Increased expression of schizophrenia-associated gene C4 leads to hypoconnectivity of prefrontal cortex and reduced social interaction. PLoS Biol. 2020 Jan 14;18(1):e3000604.

8. Hugdahl K, Sommer IE. Auditory verbal hallucinations in schizophrenia from a levels of explanation perspective. Schizophr Bull. 2018;44(2):234–41.
9. Lewandowski KE, De Paola J, Camsari GB, Cohen BM, Ongür D. Tactile, olfactory, and gustatory hallucinations in psychotic disorders: a descriptive study. Ann Acad Med Singap. 2009;38(5):383–5.
10. Mukherjee S, Shukla S, Woodle J, Rosen AM, Olarte S. Misdiagnosis of schizophrenia in bipolar patients: a multiethnic comparison. Am J Psychiatry. 1983;140(12):1571–4.
11. Maguire M, Singh J, Marson A. Epilepsy and psychosis: a practical approach. Pract Neurol. 2018;18(2):106–14.
12. Goeb JL, Even C, Nicolas G, Gohier B, Dubas F, Garré JB. Psychiatric side effects of interferon-beta in multiple sclerosis. Eur Psychiatry. 2006;21(3):186–93.
13. Lin L-R, Zhang H-L, Huang S-J, Zeng Y-L, Xi Y, Guo X-J, Liu G-L, Tong M-L, Zheng W-H, Liu L-L, Yang T-C. Psychiatric manifestations as primary symptom of neurosyphilis among HIV-negative patients. J Neuropsychiatry Clin Neurosci. 2014;26(3):233–40.
14. Narang HK. A critical review of atypical cerebellum-type Creutzfeldt-Jakob disease: its relationship to "new variant" CJD and bovine spongiform encephalopathy. Exp Biol Med (Maywood). 2001;226(7):629–39.
15. Fiorentini A, Volonteri LS, Dragogna F, Rovera C, Maffini M, Mauri MC, Altamura CA. Substance-induced psychoses: a critical review of the literature. Curr Drug Abuse Rev. 2011 Dec;4(4):228–40.
16. Tucker JA, Chandler SD, Witkiewitz K. Epidemiology of recovery from alcohol use disorder. Alcohol Res. 2020 Nov 12;40(3):02.
17. Mullen JM, Richards JR, Crawford AT. Amphetamine related psychiatric disorders. 2020 Nov 18. In: StatPearls [Internet]. Treasure Island (FL): StatPearls Publishing; 2021.
18. Nieschlag E, Vorona E. Doping with anabolic androgenic steroids (AAS): adverse effects on non-reproductive organs and functions. Rev Endocr Metab Disord. 2015;16(3):199–211.
19. Byrne A, Kirby B, Zibin T, Ensminger S. Psychiatric and neurological effects of chronic solvent abuse. Can J Psychiatr. 1991;36(10):735–8.
20. Smith CM, Komisar JR, Mourad A, Kincaid BR. COVID-19-associated brief psychotic disorder. BMJ Case Rep. 2020 Aug 11;13(8):e236940. https://doi.org/10.1136/bcr-2020-236940.
21. Pillny M, Lincoln TM. Moderne kognitive Verhaltenstherapie bei psychotischen Störungen [Modern cognitive behavior therapy of psychotic disorders]. Nervenarzt. 2020 Jan;91(1):43–9. https://doi.org/10.1007/s00115-019-00831-w.
22. Schooler NR. Relapse and rehospitalization: comparing oral and depot antipsychotics. J Clin Psychiatry. 2003;64(Suppl 16):14–7.
23. Hor K, Taylor M. Suicide and schizophrenia: a systematic review of rates and risk factors. J Psychopharmacol. 2010;24(4 Suppl):81–90.
24. Alarcón RD, Parekh A, Wainberg ML, Duarte CS, Araya R, Oquendo MA. Hispanic immigrants in the USA: social and mental health perspectives. Lancet Psychiatry. 2016;3(9):860–70.
25. Vega WA, Sribney WM, Miskimen TM, Escobar JI, Aguilar-Gaxiola S. Putative psychotic symptoms in the Mexican American population: prevalence and co-occurrence with psychiatric disorders. J Nerv Ment Dis. 2006 Jul;194(7):471–7. https://doi.org/10.1097/01.nmd.0000228500.01915.ae.
26. Anglin DM, Malaspina D. Ethnicity effects on clinical diagnoses compared to best-estimate research diagnoses in patients with psychosis: a retrospective medical chart review. J Clin Psychiatry. 2008 Jun;69(6):941–5.
27. David D, Kutcher GS, Jackson EI, Mellman TA. Psychotic symptoms in combat-related posttraumatic stress disorder. J Clin Psychiatry. 1999 Jan;60(1):29–32.
28. Escamilla M, Lee BD, Ontiveros A, Raventos H, Nicolini H, Mendoza R, Jerez A, Munoz R, Medina R, Figueroa A, Walss-Bass C, Armas R, Contreras S, Ramirez ME, Dassori A. The epsin 4 gene is associated with psychotic disorders in families of Latin American origin. Schizophr Res. 2008 Dec;106(2–3):253–7.

29. Yockey A, Stryker S. The epidemiology of cocaine use among Hispanic individuals: findings from the 2015-2018 National Survey of drug use and health. Hisp Health Care Int. 2020 Nov;24:1540415320971634.
30. Cano M, Oh S, Salas-Wright CP, Vaughn MG. Cocaine use and overdose mortality in the United States: evidence from two national data sources, 2002-2018. Drug Alcohol Depend. 2020 Sep;1(214):108148.
31. Puyat JH, Daw JR, Cunningham CM, Law MR, Wong ST, Greyson DL, Morgan SG. Racial and ethnic disparities in the use of antipsychotic medication: a systematic review and meta-analysis. Soc Psychiatry Psychiatr Epidemiol. 2013 Dec;48(12):1861–72. https://doi.org/10.1007/s00127-013-0753-4.
32. Oluwoye O, Stiles B, Monroe-DeVita M, Chwastiak L, McClellan JM, Dyck D, Cabassa LJ, McDonell MG. Racial-ethnic disparities in first-episode psychosis treatment outcomes from the RAISE-ETP study. Psychiatr Serv. 2018 Nov 1;69(11):1138–45. https://doi.org/10.1176/appi.ps.201800067.
33. Hernandez M, Hernandez MY, Lopez D, Barrio C, Gamez D, López SR. Family processes and duration of untreated psychosis among US Latinos. Early Interv Psychiatry. 2019 Dec;13(6):1389–95. https://doi.org/10.1111/eip.12779.
34. Chartier KG, Carmody T, Akhtar M, Stebbins MB, Walters ST, Warden D. Hispanic subgroups, acculturation, and substance abuse treatment outcomes. J Subst Abuse Treat. 2015 Dec;59:74–82.

Mood Disorders in Hispanic/Latinos

5

Mauricio Tohen and Hugo Gomez Rueda

Epidemiology of Depression and Bipolar Disorder in Hispanic and Latino Population

Hispanic/Latino Groups Composition

The term "*Hispanic*" refers to someone that speaks Spanish and resides in the USA, with a Latin-American ancestry [1]. The term "*Latino*" refers to the actual origin that an individual is coming from. For instance, a *Latin-American* originates from the North, Central, or South America. However, for purposes of the US Census both terms are used interchangeable [2].

Difficulties Studying Minorities, Hispanic and Latino Groups Included

In general, the inclusion of minorities in research studies represents a challenge. Certain difficulties identified while studying minority populations such as Latinos and Hispanics are: help seeking due to negative attitudes towards psychotherapy or psychiatry, language barriers, religious beliefs, lack of trust, stigma from the patient or family, fear of being reported to immigration, lack of childcare, transportation provisions, financial constraints, lack of medical insurance, the location of interview/appointments, family perspectives, the underutilization of mental health services, language used on professional interventions, lack of communication and cultural awareness between staff and participants, staff personal attributes, limited willingness and/or enthusiasm of researchers, lack of understanding the need for

M. Tohen (✉) · H. G. Rueda
Department of Psychiatry and Behavioral Sciences, University of New Mexico Health Sciences Center, Albuquerque, NM, USA
e-mail: MTohen@salud.unm.edu

R. Castilla-Puentes, T. Falcone (eds.), *Mental Health for Hispanic Communities*,
https://doi.org/10.1007/978-3-031-13195-0_5

ethnic participation, paucity of resources, appropriateness of assessment tools, non-availability of translated materials, lack of culturally competent staff, lack of culturally matched staff, underrepresentation of ethnic minorities at recruitment sites, and lack of understanding of consent process [3].

It is important to keep in mind, most of all the studies regarding minorities of Hispanics and Latinos have an inherited bias, due to the lack of or poor participation. More interventions to enhance recruitment of ethnic groups must be made in order to promote the importance of minorities in research studies, and control the aforementioned bias [3].

Prevalence of Depression on Hispanic and Latino Groups in Comparison with Other Ethnicities Living in the USA

The prevalence of depression is different among US race/ethnic groups [4]. In order to provide information regarding such differences of major depressive disorder and dysthymic disorder, the 2005 Riolo et al. article published data from The National Health and Nutrition Examination Survey III, which was conducted between the years 1988 and 1994 by the National Center for Health Statistics [4].

The sample size included 8449 subjects, between ages 15–40 years, and with 96.1% response rate in the participation. The diagnoses of major depressive disorder or dysthymic disorder were provided employing the Diagnostic and Statistical Manual of Mental Disorders, Revised Third Edition (DSM-III-R), valid at that time. The authors observed that the prevalence of major depressive disorder was higher in participants with White race and ethnicity. Moreover, Mexican Americans and White individuals had significantly earlier onset of major depressive disorder [4].

Interestingly, dysthymic disorder was more prevalent in African American and Mexican American individuals, in comparison with White participants. The lack of education, less than 8 years of education, was the risk factor associated with a higher prevalence of dysthymic disorder in the mentioned races/ethnicities, which had statistically significant p-value after the logistic regression analysis controlling for age, gender, income, education, and marital status [4].

The trend of depression in the general population has been changing over time [5]. In an annual cross-sectional study by Weinberger et al. in the year 2018 they analyzed the trend with 607,520 participants in the USA, from ages 12 years old and older. The variable analyzed was a past-year depression prevalence between the years 2005 and 2015. According to the presented trending data from the National Survey on Drug Use and Health (NSDUH), the prevalence of depression showed an increase during the studied period (Fig. 5.1) [5].

Moreover, Weinberger et al. compared depression of the surveyed years among races and ethnicities. The races and ethnicities in the study were Non-Hispanic whites, Hispanics, Non-Hispanic blacks, and all other races. The race and ethnic group that was most affected by depression in the past year of the survey was non-Hispanic whites, which was also the reference group with 95% CI (1.01–1.02), followed by all other races and ethnicities 95%CI (1.01–1.05), and Hispanics 95% CI (1.00–1.03), as shown in Fig. 5.2 [5].

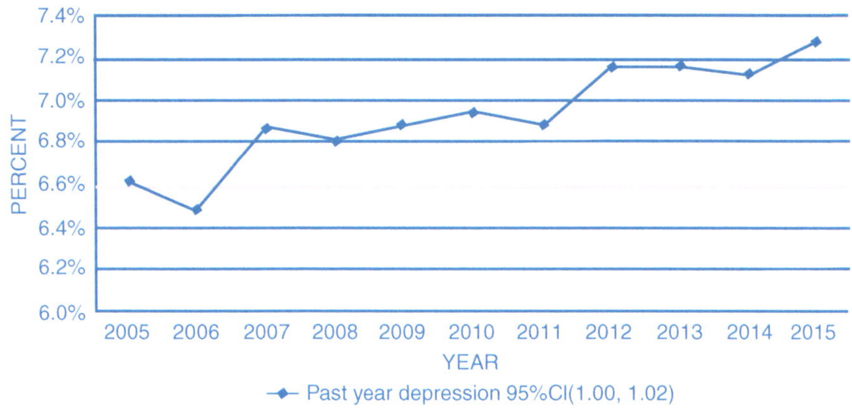

Fig. 5.1 Depression in the USA from 2005 to 2015. Reproduced with permission from [5]

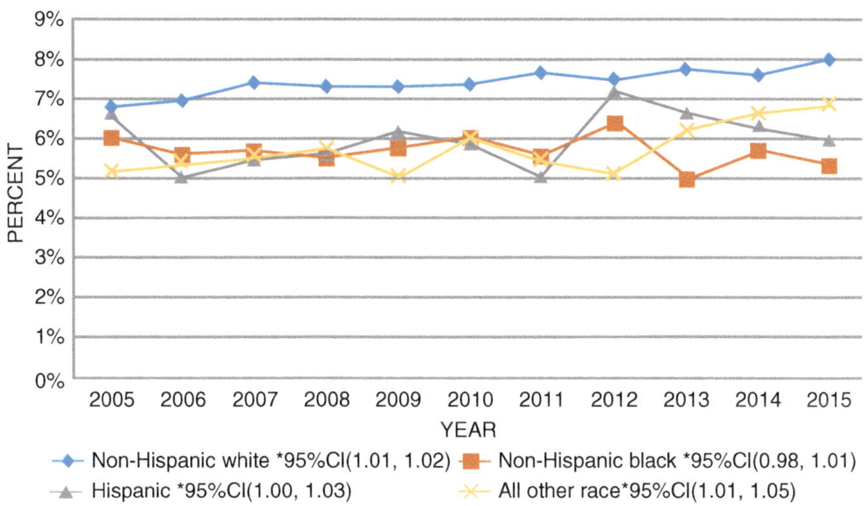

Fig. 5.2 Depression by race, 2005–2015. Reproduced with permission from [5]

Most recently, in the year 2020 Hargrove et al. showed statistical data that challenges the "immigrant paradox," as well as the prevalence showed in previous studies [6]. It is important to highlight that the authors presented results based on depressive symptoms without establishing a formal DSM-based diagnosis [6].

In order to identify depressive symptoms in 18,566 participants, the authors used four items derived from the 20-item Center for Epidemiologic Studies-Depression Scale (CES-D). The items were: (1) "you could not shake off the blues, even with help from your family and your friends," (2) "you were depressed," (3) "you were sad," and (4) "you felt happy." The participants were then enrolled in the school grades from 7 to 12 at baseline during in the years 1994–1995 (Wave I). Follow-up data were subsequently collected in the year 1996 (Wave II), 2001–2002 (Wave III), and 2008–2009 (Wave IV). The races and ethnicities taken into account in this analysis were: non-Hispanic White, non-Hispanic Black, Hispanic, and non-Hispanic Asian American. The non-Hispanic White race and ethnicity served as the reference group [6].

As shown in Fig. 5.3, African American, Hispanic, and Asian American men and women reported higher average levels of depressive symptoms across the ages of 12–42, in comparison with their White race and ethnicity counterparts. Additionally, as presented in Fig. 5.4, the presentation of such depressive symptoms occurred in early teens, with a recurrence peak in the mid and late third decade of life, extended to early forties.

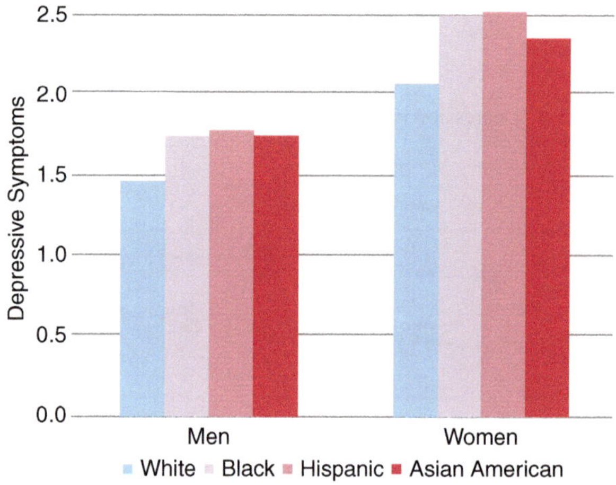

Fig. 5.3 Average depressive symptoms by race/ethnicity and gender across the ages 12–42 (N = 18,566). Reproduced with permission from [6]

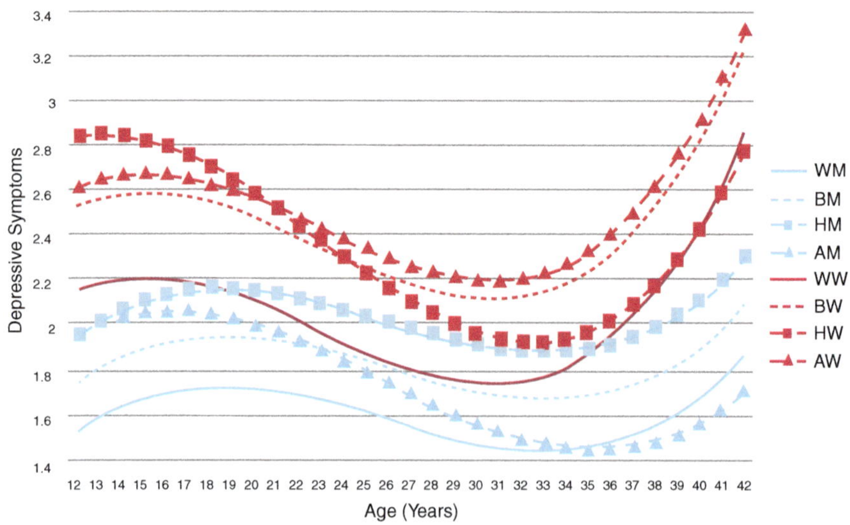

Fig. 5.4 Age trajectories of depressive symptoms by race/ethnic-gender groups across ages 12–42 (N = 18,566). Reproduced with permission from [6]

The results are highly suggestive of the existence of a bimodal presentation in depressive symptoms in the interviewed participants between ages of 12–42 years old (Fig. 5.4). Also, there is a population-dependent presentation of depressive symptoms markedly seen in minorities (Fig. 5.3). Further research is essential to determine if the results presented in the mentioned study are statistically significant, or are valid only for the studied population [6].

Prevalence of Depression on Hispanic and Latino Groups, Differences by Ethnic Group

Differences in the prevalence of depression between Hispanic and Latino groups have been identified [7, 8]. In the study in the year 2007, Alegría et al. showed differences in the prevalence reported in 2554 English- and Spanish-speaking Latinos. The population was composed by 868 Mexicans, 495 Puerto Ricans, 577 Cubans, and 614 classified as other Hispanic or Latino ethnicity and race. The analyzed data was from the National Latino and Asian American Study (NLAAS) conducted between May 2002 and December 2003. The study included an analysis of a lifetime report of psychiatric disorders, and an analysis of psychiatric disorders the previous year before the NLAAS [7].

In order to estimate psychiatric disorders rates, including depressive disorders, the instrument employed in this study was the World Health Organization Composite International Diagnostic Interview (WMH-CIDI) [9]. The variables taken into account in the analysis were gender, age, racial and ethnic group, educational level, nativity status, English-language proficiency, years of residence in the United States, age at the time of immigration, and generational status [7].

The study suggested that in comparison with Puerto Ricans men and women, Mexican men and women were less likely to have a lifetime history of depressive disorders (dysthymia or major depressive disorder), with OR = 0.57 (95% CI = 0.34–0.97) and OR = 0.69 (95% CI = 0.48–0.99) [7].

In the lifetime analysis of psychiatric disorders, it was observed that Latino men with "Excellent" or "Good English" proficiency level reported a higher risk of depression in their lifetime in comparison to Latino men with "Fair" or "Poor English" proficiency levels, with OR = 1.62 (95% CI = 1.04–2.54). This pattern was not present in the lifetime report provided by Latino women with "Excellent" or "Good English" proficiency level OR = 1.29 (95% CI = 0.96–1.72). However, in the lifetime report of psychiatric disorders, Latino women of third-generation US born were more prone to have depression in their lifetime with OR = 1.63 (95% CI = 1.02–2.62), in comparison with second generation, OR = 0.85 (95% CI = 0.58–1.23) or first generation, OR = 1 (reference group in the study) [7].

On the contrary, in the bivariate correlation in Latinos with a depressive disorder in the previous year before the NLAAS 2002–2003, there was not a significant difference between Puerto Ricans, Cubans, Mexicans, or other Latino ethnicities for women and men. Nevertheless, for Latino men with "Excellent" or "Good English" proficiency level, and Latino men of third-generation US born, there was an increased risk of having depression in the previous year before the NLAAS 2002–2003, with OR = 2.28 (95% CI = 1.14–4.58) and OR = 2.16 (95% CI = 1.20–3.89), respectively [7].

In 2014, Wassertheil-Smoller et al. presented epidemiological data from a more broad and diverse population, including Mexicans, Puerto Ricans, Dominicans, Cubans, Central Americans and South Americans in their analysis. The data was collected through the Hispanic Community Health Study/Study of Latinos (HCHS/SOL), which enrolled 16,415 Hispanic/Latino participants, between the ages of 18 and 74 years. The surveyed participants resided in communities from the Bronx, New York; San Diego, California; Miami, Florida; and Chicago, Illinois during the years 2008 to 2011 [8].

The depressive symptoms were assessed with the 10-item form obtained from the Center for Epidemiological Studies Depression Scale, CES-D10 [10]. Of the original 16,415 enrolled participants, 15,864 interviewees (97%) completed all surveys. The Odds Ratio (OR) was reported as unadjusted known as Model 1. In Model 2, the logistic regression of the OR was adjusted for age, gender, clinical site, and Hispanic/Latino background. Finally, in addition to the co-variables of Model 2, in Model 3 the logistic regression of the OR was adjusted for marital status, education, current smoker, diabetes, hypertension, dyslipidemia, obesity, and self-report of prior coronary heart disease or strokes [8].

The self-reported background of the studied cohort was represented by 40% Mexicans, 16% Puerto Ricans, 14% Cubans, 11% Central Americans, 9% Dominicans, and 7% South Americans, with an overall prevalence of depression of 27.0% (95% CI 25.8–28.2). Mexican and Puerto Rican races and ethnicities had the lowest prevalence of 22.3% (95% CI 20.4–24.3) and highest of 38.0% (95% CI 35.2–41.0) prevalence, respectively [8].

The Mexican Hispanic/Latino group was used as a reference in the study, because presented the lowest prevalence of depression. The ethnic groups presenting with more depressive symptoms were Cubans and Puerto Rican, with a higher OR, and a score of ≥10 on CES-D10, in comparison to Mexicans in the 3 proposed models. Also, for Cubans the OR was 1.36 (95% CI = 1.16–1.61) for Model 1, OR = 1.35 (95% CI = 1.04–1.75) for Model 2, and OR = 1.44 (95% CI = 1.11–1.85) for Model 3. Lastly, in the case of Puerto Ricans, the results were OR = 2.15 (95% CI = 1.82–2.54, OR = 1.62 (95% CI = 1.25–2.10), and OR = 1.54 (95% CI = 1.20–1.98) for Models 1, 2, and 3, respectively. Other aspects such as being single, or divorced, as well as second-generation immigrant, showed a significant OR with 95% CI higher than 1 in their respective comparisons in the 3 Models [8].

This study also reported the percentage of participants that were taking antidepressants for depressive symptoms. From the overall percentage in this cohort of 15,586 participants, it was reported that 5% were taking antidepressants (95% CI = 4.5–5.5). Puerto Ricans and Cubans were the ethnicities with a higher prevalence of using antidepressants, 8.3 (95% CI = 7.0–9.8) and 7.2 (5.9–8.7), respectively. On the other hand, Central Americans and South Americans had a lower prevalence of using antidepressants, with 2.5 (1.7–3.6) and 2.3 (1.4–3.6), respectively [8].

There was a significant difference in the use of antidepressants due to insurance status, with 8.2% (7.3–9.2) of insured participants in comparison with 1.8% (1.5–2.2) of uninsured participants taking antidepressants. Furthermore, among the participants taking antidepressants, 62.7% used selective serotonin reuptake inhibitors (SSRI) in monotherapy, 10.4% used tricyclic antidepressants in monotherapy, and 26.8% took a combination or other antidepressants [8].

Prevalence of Bipolar Affective Disorder on Hispanic and Latino Groups

The prevalence of bipolar affective disorder (BPAD) in minority ethnic groups has been determined by different authors. In the year 2010, Hwang et al. published a chart review analysis coming from the Bipolar Disorders Clinic seen at Stanford University between June 2000 and March 2007, of 172 patients out of the 557 outpatients. The patients had the diagnosis of Bipolar disorder type I, type II, and Not Other Specified (Unspecified according to the current DSM-5). The minority groups reported in this analysis were 86 Caucasian, 35 Latino, and 51 Asian patients. The variables taken into account in this analysis were age, gender, marital status, education, employment, insurance, and annual household income, with no statistical differences between the mentioned variables in the analyzed patients [11].

The reported prevalence was different among the studied groups. Bipolar type I disorder was more prevalent in Asian patients 58.8% (30/51) and Latinos patients 60% (21/35) compared to Caucasian patients with 37.2% (32/86) prevalence. However, bipolar II disorder and bipolar NOS were more common in Caucasian patients with 62.8% (54/86), in comparison to Latinos 40% (14/35) and Asian patients 41.2% (21/51). Moreover, in the further description of the cohort, there was no statistical difference in the type and number of medications used for treatment, the duration of illness, age of first episode, age of first treatment, and the delay in treatment. The only trending difference, with a p-value = 0.08, was in the number of visits made with a Mental Health Professional. Results indicate more visits occurred within Caucasian patients in the past 3 months of the study [11].

In the year 2017, Blanco et al. published the results of the National Epidemiologic Survey on Alcohol and Related Conditions-III (NESARC-III), conducted between the years 2012 and 2013. The sample size was 36,309 participants, ages 18 and older [12]. The results were presented as 12-month and Lifetime DSM-5 Bipolar I Disorder. The races and ethnicities in the analysis were White (used as a reference in the comparisons), Black, Native American, Asian/Pacific Islander, and Hispanic. The statistical analysis included Odds Ratio (OR), and prevalence with Standard Error (SE) [12].

It is worth noting that the reported Odds Ratio (OR) were adjusted for all other sociodemographic characteristics. In the lifetime and 12-month report in the total sample, Native American had the highest prevalence of 5.6% (SE = 1.05) and 3.9% (SE = 0.93), respectively, compared to participants with White race/ethnicity [12]. Focused specifically on Hispanic population, the authors indicated that in the

12-month and lifetime reports, the prevalence was 1.6 (SE = 0.19) and 1.9 (SE = 0.21), respectively. The adjusted OR were 0.7 (95% CI = 0.48–0.95) in the 12-month report, while in the lifetime report were 0.7 (95% CI = 0.5–0.9) [12].

Interestingly, grouping by gender, Hispanic males and females were not statistically different from their correspondent White participants. The reported 12-month prevalence in Hispanic males was 1.7% (SE = 0.25) and OR = 0.7 (95% CI = 0.47–1.05), while in Hispanic females the calculated prevalence was 1.4% (SE = 0.24), with OR = 0.7 (95% CI = 0.41–1.02). In the comparison of Hispanic and White men and women, the lifetime report showed that the prevalence was lower for Hispanics than with White participants. In their lifetime, Hispanic men participants reported a prevalence of 2.1% (SE = 0.27) with OR = 0.7 (95% CI = 0.47–1.0), and for Hispanic women the prevalence was 1.7% (SE = 0.25) with OR = 0.7 (95% CI = 0.45–0.96) [12].

Additionally, the presentation of manic symptoms of Bipolar Disorder type I, number of depressive episodes, and functional impairments have a different prevalence among the racial and ethnic groups [13]. A study made during 2001–2002 by Perron et al., presented it in the year 2010, described these differences in 43,093 interviewees, ages >15 years old, in the National Epidemiologic Survey on Alcohol and Related Conditions (NESARC). The lifetime prevalence of participants that gathered criteria for Bipolar Disorder type I was 3.6% from the surveyed population, and when allotting by race/ethnic groups the prevalence were 3.8%, 3.6%, and 3.4%, for African Americans, Whites, and Latinos, respectively, with no statistical differences [13].

Moreover, from the 16 manic symptoms included in the analysis, only 2 symptoms were statistically different among the ethnic groups. Such symptoms were excessive restlessness and inflated self-esteem/grandiosity. These symptoms were more prevalent in African Americans (39.9%) and Latinos (38.8%) in comparison with Whites (25.3%) [13]. Also, Whites had the highest rates of depressive episodes (75.0%), followed by African Americans (62.4%) and then Latinos (55.6%). The level of functioning was measured using scales from the 12-item Short-Form Health Survey, version 2 (SF-12v2). This took into account physical, mental, and psychosocial domains, where the lower scores are associated with higher functionality. In the multivariate linear analysis controlling for socio-demographics and clinical characteristics, White and African American participants had lower scores in comparison with Latinos. The means of vitality measured in SF-12v2 were 48.5 for White interviewees, and 48.8 for African American participants, with no report for Latinos [13].

Disinformation and Stigma, Two Big Barriers in the Treatment of Depression in Hispanic/Latino

The preference in the treatment of depression varies by race and ethnicity. In the year 2007, Givens et al. described these differences in different races and ethnicities. The study included 78,753 participants with significant depressive symptoms, including 68,319 Whites, 3596 African Americans, 2794 Asians/Pacific Islanders, and 3203 Hispanics. The participants with ethnic minority of African Americans,

Asians/Pacific Islanders, and Hispanics preferred counseling above antidepressant medications OR = 2.6 (95% CI = 2.4–2.8), OR = 2.5 (95% CI = 2.2–2.7), and OR = 1.8 (95% CI = 1.7–2.0), respectively. Moreover, the participants believed that antidepressants are addictive, as well as that counseling and praying were effective for the treatment of depression [14].

The agreement for an antidepressant as a primary component in the treatment of depression is different between Spanish-speaking Hispanics, English-speaking Hispanics, and non-Hispanic White individuals. During a telephone interview analysis of 839 non-Hispanic White participants and 139 Hispanic participants, Garcia et al. determined that Spanish-speaking Hispanics preferred antidepressants less with adjusted OR = 0.41 (95% CI = 0.19–0.90), while English-speaking Hispanics was OR = 1.18 (95% CI = 0.60–2.33), in comparison with White participants. The aforementioned was determined after adjusting for age, gender, history of depression diagnosis, and current depressive symptoms. Also, the biomedical explanation of depression was associated with an increased acceptance of antidepressants OR = 4.76 (95% CI = 3.13–7.14), where it can be inferred that Spanish-speaking Hispanics do not believe that depression has a biomedical explanation [15].

The stigma of Hispanic patients towards antidepressants has been consistently negative in time. In the years 2007 and 2010, Interian et al. provided statistical data that backup this asseveration [16, 17]. In the year 2007, after the analysis of qualitative data, the conclusion was that participants with Latino ethnicity believe that the use of antidepressants relates to a more severe illness, weakness or failure to cope with problems, or being under the influence of a drug [16]. In the year 2010, a more formal statistical analysis showed that Latinos who scored higher on the Stigma Concerns About Mental Health Care (SCMHC) scale, and in the Latino Scale for Antidepressant Stigma (LSAS), were less likely to take antidepressants. In participants taking medications during their participation, the OR reported for SCMHC was 0.64 (95% CI = 0.45–0.93), and LSAS was OR = 0.77 (95% CI = 0.62–0.97) [17].

Cultural Influence of Depression Manifestation in Hispanic and Latino Groups

Different Cultural Factors Influencing Depression in Hispanics

Multiple cultural components in relation to depression in Hispanic and Latino groups have been described, such as ethnic identity, familism, discrimination, and acculturative stress, among others [18]. The aspect of familism is a cultural value commonly found in Hispanic families where respect, support, obligation, and reference play a center role in the Hispanic family dynamic [19]. Familism was found to be a protective factor against depression [20–22]. However, in the year 2016, Valdvieso-Mora et al. conducted a systematic review of 39 studies related to familism and depression, suicide, internalizing symptoms, externalizing symptoms, and substance abuse. The results showed that the effect size measured with a *Cohen's d* for depression in patients with familism was between $d = 0.02$ and 0.75. Moreover, out of the 24 studies included for depression and familism, 13 studies

(54.2%) showed no effect of familism in depression, 8 studies (33.3%) presented a small effect, and 3 studies (12.5%) indicated a moderate effect [19].

In order to determine the role in the development of depression in Hispanic/Latino groups, other cultural aspects have been studied. In the year 2019, Calzada et al. studied how depression influence the pressure to acculturate, the pressure against acculturation, perceived discrimination, ethnicity identity, and familism, within 175 Hispanic mothers. Surprisingly, the results showed that strong ethnic identity and perceived discrimination were protective factors for depression later in life in Hispanic mothers. The logistic regression coefficients were − 0.17 (95% CI = −0.32; −0.01), and − 0.21 (95% CI = −0.39; −0.03), respectively. Interestingly, the aspects of pressure to acculturate and familism were identified as a risk factor for depression later in life in the same population, with logistic regression coefficients of 0.14 (95% CI = 0.01–0.27), and 0.13 (95% CI = 0.00–0.25) [18].

Other aspect that many Hispanics and Latinos face when residing in the USA is discrimination [23]. In the year 2014, Budwhani et al. explored the impact of nativity and discrimination, through 5 logistic regression Models of race, ethnicity, nativity, gender, socioeconomics, health lifestyles, discrimination, and race/nativity interactions. The study was conducted with 17,249 participants of the Collaborative Psychiatric Epidemiology Surveys. Participants included were 3369 Hispanics (19.53%), whom 1920 of them (11.13%) were foreign-born. Surprisingly, for Hispanics being foreign-born did not represent a risk factor to develop depression, OR = 0.946 (95% CI = 0.626–1.429), taking into account their Model 5, which included an interaction term of born origin with race/ethnicity. However, discrimination was a factor that can develop depression OR = 1.446 (95% CI = 1.374–1.523, p-value <0.001) in their Model 4 adding the discrimination scale. The scale of discrimination included aspects such as felt dishonest, treated as less smart, treated with disrespect, threatened, or called names [23].

It is worth noting certain factors that Latino immigrants will face, or are facing, during the acculturative process which contribute to added stress. These factors include: 1) *Environmental* (financial, language barrier, lack of access to health care, unsafe neighborhoods, unemployment, lack of education), 2) *Social/Interpersonal* (loss of social networks, loss of social status, family conflict, intergenerational conflicts, changing gender roles), and 3) *Societal* (discrimination, stigma, legal status, political, and historical forces) [24]. Such factors can impact the mental health of the Latino and Hispanic patient.

Psychopharmacogenetics and Genetic Studies in Hispanic and Latinos Populations

Psychopharmacogenetic Studies

The number of studies related to pharmacogenomics in Hispanic populations is significantly smaller in comparison with Caucasian and Asian populations [25]. The term psychopharmacogenetics is arbitrarily proposed by the authors to indicate the association between genetics and psychopharmacology. It is also a fundamental field for our better understanding of the observed differences in response to treatment in psychotropics in different ethnicities.

The association between Single Nucleotide Polymorphisms (SNPs) in genes related to major depression disorder (MDD), and antidepressant response, has been studied in Mexican Americans [26]. Dong et al. sequenced the following genes: ATP-binding cassette subfamily B member 1 (ABCB1), the noradrenaline, dopamine, three serotonin transporters (SLC6A2, SLC6A3, and SLC6A4), cyclic AMP-responsive element binding protein 1 (CREB1), corticotropin-releasing hormone receptor 1 (CRHR1), and neurotrophic tyrosine kinase type 2 receptor (NTRK2). The Mexican American population included 536 random participants originating from Los Angeles, CA in which 264 participants were healthy controls and 272 participants had MDD. An extensive analysis comprised a correction based on a general linear model, controlling for age, gender, and baseline HAM-D21 score, and then applying a Benjamini–Hochberg false discovery rate after correcting for multiple testing. After an 8-week antidepressant treatment with desipramine, in Mexican Americans population of Los Angeles, the polymorphisms rs2289657 and rs56142442 who were homozygous for C allele, synonymous of the gene NTRK2, were associated with higher relative reduction of HAM-D21 scale score (27% larger), pointing at least to a distinction in that specific population [26].

Other studies had centered their attention on the cytochrome P450, because it is a limiting factor for drug availability and effectivity. It metabolizes between 70% and 80% of all drugs in clinical use [27]. From that family, the subunit 2D6 (CYP2D6) is very important to psychopharmacogenetics, because, according to *National Institutes of Health* (NIH), CYP2D6 metabolizes approximately 25% of the prescribed drug, including antipsychotics, antidepressants, beta-blockers, analgesics, antiarrhythmics, and antiemetics [28]. Implicating that a Single Nucleotide Polymorphism (SNP), or a Copy Number Variation (CNV) in CYP2D6, could affect the therapeutic response in psychotropic medications [27].

In 2017, Salyakina et al. conducted a study in South Florida to measure the allele frequencies of CYP2D6, including 22 different alleles. The population was 413 participants, including 75% self-reported as Hispanics. The study showed statistical differences between White Hispanics and Black Hispanics in the alleles *4, *17, *41, and *2A in the gene CYP2D6, which indicates a distinctive Hispanic population in South Florida. However, the frequencies between poor, normal, intermediate and ultra-rapid metabolizers were not different between the different racial and ethnic groups, and this effect can be associated with the sample size [29].

Other Genetic Studies in Hispanic and Latino Population

In an effort to identify distinctive SNPs in the Hispanic population with depressive symptoms, Dunn et al. explored this possibility in 12,310 adults from the Hispanic Community Health Study/Study of Latinos (HCHS/SOL) which was the most extensive study of Genome-wide Association Study (GWAS) made with Latinos. The participants had reported depressive symptoms in the past week before their interview. The phenotypes taken into account in the analysis were: total depression score, total score modified to account for psychiatric medication use, and score excluding antidepressant medication users. Interestingly, no single SNP was an indicator of depression in association with the three mentioned phenotypes. Moreover, there was no genetic evidence of risk factors across ancestry groups [30].

Other GWAS associations have been found with depressive symptoms in Hispanic/Latina women. In the year 2015, Dunn et al. analyzed the SHARe cohort of the Women's Health Initiative, including 3138 Hispanics/Latina women. Although no SNPs achieved genome-wide significance, the analysis showed top signals associated with MDD. The signals associated to depression in Hispanics/Latina women were rs2532087 (located 27 kb from the gene CD38, $p = 2.44 \times 10^{-7}$) and rs4542757 (intronic to the gene DCC (deleted in colon cancer), $p = 7.31 \times 10^{-7}$) [31].

Similarly to other diseases, MDD is present in different subpopulations that have shared genetic similarities, as well as certain clinical traits [32]. In the year 2017, Yu et al. grouped 203 Mexican American with MDD and 196 healthy controls aged 19–65 years, based on their SNP characteristic using 83,898 SNP variants. From that analysis, a total of 19 genes were significantly associated with MDD with the genome-wide false discovery rate < 0.05. The clustering process employed the tree method and the Hamming distance matrix. During their clustering analysis, the authors identified that 41 patients with MDD created a unique cluster. This cluster of patients was grouped using the 83,898 SNPs and also using the 19 statistically significant genes. The authors named this group as the latent subtype of the Mexican American patients with MDD. This latent subtype population had statistically higher levels of anxiety and fewer symptoms of middle insomnia, depersonalization, derealization, and paranoid symptoms in the HAM-D of 21 items [32].

In bipolar disorder, there is a possible association between mood oscillations of bipolar affective disorder (BPAD), and the rhythmic change in expression of circadian genes in Latino population. Gonzalez et al. analyzed 884 individuals from 207 pedigrees (473 BPAD phenotype and 411 unaffected family members). Although none of the SNPs were significantly associated with BPAD after correction for multiple testing, the permutation analysis showed that a 4-locus with the SNP rs1534891 in the gene CSNK1E, and a 3-locus haplotype (GAA) of the gene ARNTL, had a statistically significant association with BPAD. It is essential to highlight the importance of a confirmatory study in a larger Latino population with BPAD to corroborate this observation [33].

References

1. United State Census Hispanic Origin. https://www.census.gov/topics/population/hispanic-origin.html. Accessed 14 May 2020.
2. US Census Bureau About Hispanic Origin. https://www.census.gov/topics/population/hispanic-origin/about.html. Accessed 18 Feb 2020.
3. Brown G, Marshall M, Bower P, Woodham A, Waheed W. Barriers to recruiting ethnic minorities to mental health research: a systematic review. Int J Methods Psychiatr Res. 2014;23:36–48.

4. Riolo SA, Nguyen TA, Greden JF, King CA. Prevalence of depression by race/ethnicity: findings from the National Health and nutrition examination survey III. Am J Public Health. 2005;95:998–1000.
5. Weinberger AH, Gbedemah M, Martinez AM, Nash D, Galea S, Goodwin RD. Trends in depression prevalence in the USA from 2005 to 2015: widening disparities in vulnerable groups. Psychol Med. 2018;48:1308–15.
6. Hargrove TW, Halpern CT, Gaydosh L, Hussey JM, Whitsel EA, Dole N, Hummer RA, Harris KM. Race/ethnicity, gender, and trajectories of depressive symptoms across early- and mid-life among the add health cohort. J Racial Ethn Heal Disparities. 2020:1–11.
7. Alegría M, Mulvaney-Day N, Torres M, Polo A, Cao Z, Canino G. Prevalence of psychiatric disorders across Latino subgroups in the United States. Am J Public Health. 2007;97:68–75.
8. Wassertheil-Smoller S, Arredondo EM, Cai J, et al. Depression, anxiety, antidepressant use, and cardiovascular disease among Hispanic men and women of different national backgrounds: results from the Hispanic community health study/study of Latinos. Ann Epidemiol. 2014;24:822–30.
9. Kessler RC, Ustün TB. The world mental health (WMH) survey initiative version of the World Health Organization (WHO) composite international diagnostic interview (CIDI). Int J Methods Psychiatr Res. 2004;13:93–121.
10. Andresen EM, Malmgren JA, Carter WB, Patrick DL. Screening for depression in well older adults: evaluation of a short form of the CES-D (Center for Epidemiologic Studies Depression Scale). Am J Prev Med. 1994;10:77–84.
11. Hwang SHJ, Childers ME, Wang PW, Nam JY, Keller KL, Hill SJ, Ketter TA. Higher prevalence of bipolar i disorder among Asian and Latino compared to Caucasian patients receiving treatment. Asia Pac Psychiatry. 2010;2:156–65.
12. Blanco C, Compton WM, Saha TD, Goldstein BI, Ruan WJ, Huang B, Grant BF. Epidemiology of DSM-5 bipolar I disorder: results from the National Epidemiologic Survey on alcohol and related conditions–III. J Psychiatr Res. 2017;84:310–7.
13. Perron BE, Fries LE, Kilbourne AM, Vaughn MG, Bauer MS. Racial/ethnic group differences in bipolar symptomatology in a community sample of persons with bipolar I disorder. J Nerv Ment Dis. 2010;198:16–21.
14. Givens JL, Houston TK, Van Voorhees BW, Ford DE, Cooper LA. Ethnicity and preferences for depression treatment. Gen Hosp Psychiatry. 2007;29:182–91.
15. Fernandez Y, Garcia E, Franks P, Jerant A, Bell RA, Kravitz RL. Depression treatment preferences of Hispanic individuals: exploring the influence of ethnicity, language, and explanatory models. J Am Board Fam Med. 2011;24:39–50.
16. Interian A, Martinez IE, Guarnaccia PJ, Vega WA, Escobar JI. A qualitative analysis of the perception of stigma among Latinos receiving antidepressants. Psychiatr Serv. 2007;58:1591–4.
17. Interian A, Ang A, Gara MA, Link BG, Rodriguez MA, Vega WA. Stigma and depression treatment utilization among Latinos: utility of four stigma measures. Psychiatr Serv. 2010;61:373–9.
18. Calzada EJ, Sales A. Depression among Mexican-origin mothers: exploring the immigrant paradox. Cult Divers Ethn Minor Psychol. 2019;25:288–98.
19. Valdivieso-Mora E, Peet CL, Garnier-Villarreal M, Salazar-Villanea M, Johnson DK. A systematic review of the relationship between familism and mental health outcomes in latino population. Front Psychol. 2016; https://doi.org/10.3389/fpsyg.2016.01632.
20. Calzada EJ, Tamis-LeMonda CS, Yoshikawa H. *Familismo* in Mexican and Dominican families from low-income, urban communities. J Fam Issues. 2013;34:1696–724.
21. Umaña-Taylor AJ, Updegraff KA, Gonzales-Backen MA. Mexican-origin adolescent mothers' stressors and psychosocial functioning: examining ethnic identity affirmation and Familism as moderators. J Youth Adolesc. 2011;40:140–57.
22. Zeiders KH, Updegraff KA, Umaña-Taylor AJ, Wheeler LA, Perez-Brena NJ, Rodríguez SA. Mexican-origin youths' trajectories of depressive symptoms: the role of familism values. J Adolesc Health. 2013;53:648–54.
23. Budhwani H, Hearld KR, Chavez-Yenter D. Depression in racial and ethnic minorities: the impact of nativity and discrimination. J racial Ethn Heal disparities. 2015;2:34–42.

24. Caplan S. Latinos, acculturation, and acculturative stress: a dimensional concept analysis. Policy Polit Nurs Pract. 2007;8:93–106.
25. Barron C, Tonarelli S, Delozier A, Briones D, Su B, Rubin L, Chun X. Pharmacogenetics of antidepressants, a review of significant genetic variants in different populations. Clin Depresion. 2016;2:1000109.
26. Dong C, Wong ML, Licinio J. Sequence variations of ABCB1, SLC6A2, SLC6A3, SLC6A4, CREB1, CRHR1 and NTRK2: association with major depression and antidepressant response in Mexican-Americans. Mol Psychiatry. 2009;14:1105–18.
27. Zanger UM, Schwab M. Cytochrome P450 enzymes in drug metabolism: regulation of gene expression, enzyme activities, and impact of genetic variation. Pharmacol Ther. 2013;138:103–41.
28. CYP2D6 gene–Genetics Home Reference–NIH. https://ghr.nlm.nih.gov/gene/CYP2D6. Accessed 29 Feb 2020.
29. Salyakina D, Roy S, Wang W, et al. Results and challenges of cytochrome P450 2D6 (CYP2D6) testing in an ethnically diverse South Florida population. Mol Genet Genomic Med. 2019; https://doi.org/10.1002/mgg3.922.
30. Dunn EC, Sofer T, Wang MJ, et al. Genome-wide association study of depressive symptoms in the Hispanic community health study/study of Latinos. J Psychiatr Res. 2018;99:167–76.
31. Dunn EC, Wiste A, Radmanesh F, et al. Genome-wide association study (GWAS) and genome-wide by environment interaction study (GWEIS) of depressive symptoms in African American and Hispanic/Latina women. Depress Anxiety. 2016;33:265–80.
32. Yu C, Arcos-Burgos M, Licinio J, Wong ML. A latent genetic subtype of major depression identified by whole-exome genotyping data in a Mexican-American cohort. Transl Psychiatry. 2017;7:–e1134.
33. Gonzalez R, Gonzalez S, Villa E, et al. Identification of circadian gene variants in bipolar disorder in Latino populations. J Affect Disord. 2015;186:367–75.

Anxiety in Hispanics

6

Pilar Lachhwani

Anxiety Disorders in Hispanics

As a Hispanic myself, I wrote this chapter to reconcile my understanding of anxiety presentation in the Hispanic population.

I organized this chapter by using the same framework of diagnostic entities established by the Diagnostic and Statistical Manual of Mental Health Disorders (DSM-5). The reason I did this is because the DSM-5 diagnostic criteria is most familiar to the US practitioner.

As health care providers, understanding cultural context, barriers, and singularities of Hispanics will give us a higher leverage point in the clinical setting to increase diagnostic accuracy and patient engagement. This will translate to provide comprehensive, compassionate, and effective care for the Hispanic/Latino population.

In order to give practitioners a deeper understanding of the cultural background of their Hispanic patients, I will provide general information about Hispanics to complement the DSM-5 diagnostic criteria.

There is merit in complementing a diagnostic formulation with a cultural framework. Such an approach can enrich a patient's and a provider's experience.

A study published in 1999 [1] suggested that essential components would include:

(a) Cultural identity of the individual.
(b) Cultural explanation of the individual's illness.
(c) Cultural factors related to psychosocial environment and functioning.
(d) Cultural elements of the relationship between the individual and the clinician.
(e) Cultural assessment for comprehensive diagnosis and care.

P. Lachhwani (✉)
Department of Psychiatry, Cleveland Clinic, Cleveland, OH, USA
e-mail: lachhwp@ccf.org

R. Castilla-Puentes, T. Falcone (eds.), *Mental Health for Hispanic Communities*,
https://doi.org/10.1007/978-3-031-13195-0_6

Who Are the Hispanic/Latinos in the United States?

Hispanic and Latino Americans are people who are descendant from Iberia (Spain, Portugal), and Latin American countries.

The U.S. Census Bureau defines being Hispanic/ Latino as an ethnicity, rather than a race, and thus people of this group may be of any race, and their origins trace back to Africa, Asia, Europe, or the Americas.

The term Hispanics/Latinos will be used in this chapter.

Demographics

After Asian Americans, Hispanics are the second fastest growing ethnic group in the United States. As of 2019, the Census Bureau estimated that there were almost 60 million Hispanics living in the United States (about 18% of the overall population) and by 2050 is predicted to increase to 97 million (one fourth of the US projected population on that date) [2].

2019 Census Bureau found that 62% of the nation's Hispanic population were of Mexican origin, 9% were from Puerto Rico, 4% were Cubans, and 9.3% were Central Americans.

The remaining Hispanic population were from other Central and South American countries or from Spain [2].

Central Americans are the newest Latino subgroup in the United States due to political and civil wars in El Salvador, Guatemala, and Nicaragua. These conflicts resulted in significant emigration from these countries during the 1980s. After hurricane Mitch in Honduras and Nicaragua in 1998, and earthquakes in El Salvador in early 2001, the United States designated nationals from those countries eligible for Temporary Protected Status (TPS). Proposed changes to not renew the TPS for those countries in 2017–2018 are been challenged in federal court in 2020 [3].

Education

Hispanics compromised 24% of all enrollments in the United States in 2011–2012, including 52% and 51% of enrollment in California and Texas, respectively. However, academic achievement in early childhood, elementary and secondary education of this subset of the population lags behind other groups. Latinos have the highest dropout rate of any group. Only 52.4% of Latinos over the age of 25 have graduated from high school, and only 10.4% have graduated from college [4].

Overall, Hispanics have less formal education than the national average. There are several potential reasons for the education disparities including language barriers, poverty, and immigration status.

Immigration Status

The experiences Hispanics have upon entering the United States vary greatly. The U.S. Government has provided Cubans support through refugee or reentrant status, work permits, and citizenship, due to their fleeing an oppressive Communist Government [4].

Fifty one percent of Cuban immigrants have become US citizens. Puerto Ricans are by definition US citizens and, as a result, have access to government sponsored support services.

However, Central American immigrants, predominantly from El Salvador, Honduras, and Guatemala are not recognized as political refugees. The 2018 proposed changes to immigration policies include zero tolerance and criminalization of illegal entry. Often this leads to children being separated from their parents.

Furthermore, the public-charge rule proposes additional barriers for immigrants who apply for visas and green cards, changes to Deferred Action for Childhood Arrivals (DACA), the use of Medicaid, the Children's health insurance program (CHIP), multiple housing programs, the supplemental nutrition assistance program (SNAP), and temporary protective status (TPS). Hispanics depend on these programs greatly to survive.

Changes to these programs could have extremely negative implications on the mental and physical health of Hispanic children, adults, and family functioning [5].

Health Care

In 2010, 43.25% of working-age Hispanic adults did not have health insurance. After the passage of the affordable care act (ACA) in 2010, the percentage of uninsured Hispanics has since fallen to 24.8% in 2016 [6].

While ACA has been able to help documented immigrants, it fails to help undocumented immigrants or legal immigrants with less than five years' residence in the United States to gain coverage.

Income

The current income levels of the Latino subgroups are related to the political and historical circumstances of their immigration as well as their educational status.

Elite Cuban immigrants have contributed in part to the relatively strong economic status of Cuban Americans in contrast to Mexican Americans, Puerto Ricans, and Central Americans, most of whom come to the United States as unskilled laborers.

According to the U.S. Census, the poverty rate for Hispanics was 18.3% in 2017, down from 19.4% in 2016. Hispanics accounted for 10.8 million individuals in poverty. In comparison, the average poverty rate in 2017 for non-Hispanic white Americans was 8.7% [7].

Language

The Spanish language is a unifying factor among Hispanics and is tied with the person's family, heritage, and overall culture. Spanish is the predominant language spoken at home in most Hispanic families. About 40% of Hispanics reported that they did not speak English at all or did not speak well enough to communicate effectively [8]. Needless to say, many Hispanics feel more comfortable speaking in Spanish than in English.

Religion

Among American Hispanics, as of 2018–2019, 47% are Catholic, 24% are Protestant, 1% are Mormon, fewer than 1% are Orthodox Christian, 3% are members of non-Christian faiths, and 23% are unaffiliated. The proportion of Hispanics who are Catholic has dropped from 2009 (was 57%) while the proportion of unaffiliated Hispanics has increased since 2009 (was 15%) [9].

Media

Radio is the largest non-English broadcasting media used by many Hispanics. Spanish language media serves to promote a sense of group consciousness among Latinos by reinforcing roots in Latin America.

Family Values

Family is very important for Hispanics.

Hispanic families prefer to live near other family members. The role of grandparents is very important in the upbringing of children. The extended family plays an important role and frequent social, family gatherings are common including birthdays, quinceaneras (15th), graduations, weddings, etc.

Cultural Matters

The geographic, political, social, economic, and racial diversity of Hispanic and Latino Americans makes all Hispanics very different. Despite this, language, religion, media, cuisine, and family traditions tend to unite Hispanics from diverse backgrounds.

Marin and Marin (1991) described key characteristics of Hispanic culture with respect to the following values [10]:

(a) Group orientation rather than individualistic orientation.
(b) Emphasis on harmony in interpersonal relations and the avoidance of confrontation.
(c) Strong loyalty and attachment to one's nuclear and extended family.
(d) Deference and obedience to authority figures and revered relatives.
(e) Preference for closeness in interpersonal space, thus perceiving physical and social space as aspects of interpersonal warmth.
(f) Present time orientation focused on here-and-now activities rather than punctuality or a detailed planning of future activities.
(g) Strong gender role expectations for men and women.

Anxiety General Considerations

In the United States, anxiety disorders have a lifetime prevalence of about 46.4%, followed by mood disorders and somatoform disorders [11].

Higher rates of anxiety disorders are found in Hispanics either born in the United States or those who have lived in the country for more than thirteen years [12].

There are many factors contributing to the high prevalence of anxiety in the Hispanic population. Some of these factors include family dysfunction through separation, deportation, larger and crowded households, low educational and economic status.

Hispanic females tend to be more expressive of internalized emotional distress, particularly through somatic expressions, while undergoing acculturation [13–15].

Some other studies indicated Mexican-American women over age 40 and people from Puerto Rico had higher rates of somatic symptoms than other non-Hispanic populations [14].

Glover and colleagues [13] found that Hispanic children in middle schools, specifically Mexican-origin youth from Texas, reported more anxiety-related problem behaviors than white students. Mexican-American adults and adolescents in general have been reported to have high levels of anxiety. Factors associated with higher anxiety scores included being born outside USA, linguistic fluency, an absent father, uneducated mothers, and household size and acculturation difficulties.

The Puerto Rican population seems to have a similar risk for psychiatric disorders when compared to the US population. This may be explained by the fact that Puerto Rico has been under the influence of US culture for more than a century and has adopted many aspects of the US culture and lifestyle. There is frequent circular migration between the island and the USA, which may also explain the lack of differences [8].

Diagnostic and Testing Issues

About 40% of Hispanics reported that they did not speak English at all or did not speak well enough to communicate effectively. Furthermore, there is no national data to indicate the language skills of the mental health professionals. A large number of English-speaking only providers are evaluating bilingual children and adults with limited English proficiency. This could affect the quality of the evaluation in many aspects [16].

Several studies have found that bilingual patients are evaluated differently when interviewed in English as opposed to Spanish [17]; however, the extent to which these factors result in misdiagnosis is not known. Further research is needed to clarify how cultural and linguistic factors influence diagnosis.

Summary of Anxiety Disorder measures most Commonly Used

Measure	Construct Assessed	Available in Spanish	Recommended Use
Beck anxiety inventory (BAI)	Consists of 21 items measures primarily somatic symptoms	YES	In English/Spanish Strong psychometrics Highly recommended for Hispanics
Anxiety severity index (ASI)	Measures both psych and somatic symptoms	YES	In English/Spanish Good psychometrics Recommended for Hispanics
State-trait anxiety inventory (STAI)	Examines two dimensions: State and trait anxiety	YES	English and Spanish version (IDARE) good psychometrics, recommended for use in Hispanics
PTSD checklist (PCL)	Assess 17 symptoms associated with PTSD	YES	Both the English and Spanish version are recommended for use with Hispanic/Latinos
Structured clinical interview for DSM-IV Axis I (SCID-I)	Assess the presence of DSM-IV anxiety dis.	YES	It has not been standardized on Hispanic population, bilingual interviewers use it
GAD-7	Assess psych symptoms	YES	Self-administered, widely used, not-validated for Hispanics
GAD Q-IV	Full diagnostic criteria	NO	Good psychometric properties, not validated for use in Hispanics
Anxiety disorder interview schedule (ADIS-IV)	Assess anxiety and mood disorders	NO	Good psychometric properties, not validated for use in Hispanics

Anxiety Disorders per Diagnostic and Statistical Manual of Mental Disorders (DSM-5)

Topics most relevant to Hispanics are listed first:

1. Other specified anxiety disorder including cultural concepts of distress related to Hispanics. In the appendix of DSM-5: Ataque de Nervios, Nervios, and susto.
2. Problems related to the social environment including:
   (a) Acculturation difficulty.
   (b) Social exclusion or rejection.
   (c) Target of (perceived) adverse discrimination or persecution (immigration stress).
3. Separation anxiety.
4. Selective mutism.
5. Specific phobia.
6. Social anxiety disorder.
7. Panic disorder-panic attack specifier.
8. Agoraphobia.
9. Generalized anxiety disorder.
10. Substance/Medication-induced anxiety disorder.
11. Anxiety disorder due to another medical condition.
12. Anxiety that is better explained by another mental condition.
13. Other relevant anxiety topics to consider.

Other Specified Anxiety Disorder Including Cultural Concepts of Distress Related to Hispanics in the Appendix of DSM-5

Ataque de Nervios, Nervios, and Susto

Also, notice: mal de ojo (evil eye, jealousy, envy), Not included in the DSM.

The presentation of symptoms in this category cause significant distress but fall short of meeting the full criteria for any of the anxiety disorder class.

DSM diagnostic criteria prioritizes psychological over somatic symptoms of anxiety and may not capture anxiety that does not conform to these diagnostic assumptions. Cultural syndromes help to expand beyond usual mental health criteria [1].

Ataque de Nervios (Attack of Nerves)

It is common among Latinos, characterized by being out of control with intense emotional upset, acute anxiety, anger, or grief; screaming, crying, trembling, and sometimes verbal and physical aggression. It is commonly associated with Caribbean Latinos [18].

Other symptoms might include:

Dissociative experiences (e.g., depersonalization, derealization, amnesia).

Non-epileptic seizures, fainting episodes, suicidal behaviors.

"Ataque de Nervios" is related to stressful family events. Events can include news of a relatives' death, conflicts with a spouse or children, or witnessing an accident involving a family member. "Ataque de Nervios" might represent normative expressions of acute distress (e.g., at a funeral) without clinical sequelae.

"Ataque de Nervios" can overlap with panic attacks, panic disorders, dissociative disorders, conversion disorders, intermittent explosive disorder, trauma and stressor related disorder.

"Ataque de Nervios"in Caribbean Latinos, Puerto Ricans, and Dominicans may meet criteria for panic disorders.

Nervios (Nerves)

Emotional distress among Latinos is expressed as "nervios." It is a general state of vulnerability to stressful life experiences, and includes somatic disturbances, and inability to function.

Somatic symptoms attributed to "nervios" are headaches, irritability, stomach disturbances, sleep difficulties, nervousness, tearfulness, lack of concentration, trembling, tingling sensations, and dizziness ("mareos").

"Nervios" share many commonalities with adjustment, anxiety, depressive, dissociative, somatic symptom, or psychotic disorders.

Many people of Mexican origin loosely apply the concept of "Nervios" to distress that is not associated with DSM disorders, as well as to distress that is associated with anxiety and depression [19].

Susto (Fright)

Susto is an expression of tribulation and tragedy, common among Latinos in the United States and Latin America.

Symptoms reported include poor sleep, appetite disturbances, low self-esteem, sadness, interpersonal sensitivity, and lack of motivation to do anything.

Somatic symptoms may include headaches, muscle aches and pains, cold in the extremities, stomach aches, and diarrhea.

"Susto" may overlap with major depressive disorder, post-traumatic stress disorder, other specified or unspecified trauma and stressor-related disorders, and somatic symptom disorder.

Problems Related to the Social Environment Relevant to Hispanic Minorities in the United States Include

(a) Acculturation difficulty.
(b) Social exclusion or rejection.
(c) Target of (perceived) adverse discrimination or persecution (immigration stress).

(a) *Acculturation Difficulty.*

Consider when an individual has difficulty in adjusting to a new culture and is not functioning well.

Multiple aspects of acculturation could be related to higher rate of mental disorders like anxiety in Hispanics, for instance, changes in cultural values and practices, the stressors associated with it or negative encounters with American Institutions (e.g., employers, schools) would be some examples [20].

(b) *Social Exclusion or Rejection.*

Consider when there is an imbalance of social power resulting in social exclusion or rejection by others. Many Hispanics experience bullying, teasing, and intimidation by others; verbal abuse and humiliation; exclusion from the activities of peers, workmates in their social environment.

Poverty and inequality by themselves foster exclusion in general populations.

Many Hispanic families live in low-income neighborhoods prone to violence and gang activities. They live in constant fear which can contribute to development of anxiety, PTSD, aggression, and depression.

When referring to exclusion or rejection, consider that many Hispanic immigrants come to the United States as unskilled laborers. They lack social and economic resources to ease their adjustment to the new country. They may have to relocate frequently to meet job demands, which strains family ties and makes developing friendships difficult. This is a hurdle in developing a sense of belonging to a determined area or community.

(c) *Target of (Perceived) Adverse Discrimination or Persecution.*

This is when there is perceived or experienced discrimination against a person based on his or her ethnicity, race, gender, gender identity, religion, or sexual orientation.

Studies reported that 31% of Hispanics reported personal experiences with discrimination while 82% believed that discrimination plays a crucial role in whether they will find success living in the USA. The current legislation on immigration policies plays a crucial role in creating a hostile and discriminatory environment for immigrants. The immigrant experience is associated with lower self-esteem, internalizing symptoms, and problematic behaviors among Mexican youth.

It is also known that more time spent living in the USA is associated with increased feelings of distress, depression, and anxiety [12].

Like many other Hispanic and Latin American groups that migrate to the USA, all immigrant groups are often stigmatized and given a bad reputation.

Separation Anxiety Disorder

Separation anxiety is when an individual is afraid of being separated from an attached person including a pet. The person develops extreme anxiety as a result from separation. While separation anxiety is a normal stage of development for infants and toddlers, most children outgrow it by 3 years of age. Childhood adversity and traumatic events increase the risk of intense and/or prolonged separation anxiety. When separation anxiety occurs in teenagers and adults, it happens in the context of moving out or going away, leaving home for college, going to work, illness or death of a loved one, and divorce of parents. This condition also affects new mothers when temporarily separated from their newborn, especially in cases of high risk and stressful pregnancies.

Adults and children differ in their somatic symptoms. Symptoms that are usually prominent in children such as nausea and stomach aches, seem to be less frequent in adults who instead show more cognitive and emotional symptoms (making frequent calls, adhering to rigid routines, talking excessively).

Cultural Issues

In some cultures, children stay with parents for a long time. There is wide variation across countries with respect to the age at which children are expected to leave their parents' home. Hispanics tend to have large multigenerational families and place a high value on interdependence among family members [10].

Hispanic families might experience separation during the process of migrating to the United States. Family separation puts US-born children, undocumented children and their undocumented parents at high risk for separation anxiety, depression, PTSD, and family maladaptive dynamics.

One study found that 85% of Central American youth were separated from both parents during immigration. If separation is from only one parent, it is most likely to be from the father. About 28% of the children have also been separated from their siblings [21].

Children and adults will experience emotional trauma with separation, often resulting in long-term changes in behaviors. Harsh immigration laws and re-entry restrictions make reunification difficult which further affect the mental health of children and parents.

Hispanic children were more likely to display a primary diagnosis of separation anxiety and were more likely to come from families with lower incomes compared to Caucasians [22].

Epidemiology

Separation anxiety is among the most common type of anxiety disorder in children.

It accounts for half of all referrals for mental health treatment for anxiety in children [23].

In children ages 7–11, 4% have the disorder.

In adolescents, the 12-month prevalence is 1.6%.

In adults in the United States is 0.9%–1.9%.

The disorder is equally common in males and females.

Data from the World Health Organization (WHO) shows lifetime prevalence of separation anxiety disorder as high as 4.8% across countries occurring after the age of 18.

Diagnosis

A complete psychiatric/psychological evaluation is needed when symptoms of separation anxiety are excessive for the developmental age and interfere with functioning. A suggested screening tool for children is the Children's separation anxiety symptom inventory and for adults is the adult separation anxiety symptom questionnaire (ASA-27). These surveys have not been validated for Hispanics.

Treatment

A multimodal comprehensive treatment approach may include cognitive behavioral therapy (CBT), family education, family psychosocial interventions, and pharmacological interventions.

Most relevant data to treat separation anxiety is reported to have a trial of CBT first if available.

Coping Cat is the most widely used form of CBT for children to recognize anxious feelings and physical reactions to anxiety and develop a plan to face and manage fears about separation and uncertainty.

Selective Serotonin Reuptake Inhibitors (SSRIs) may be prescribed in cases when the symptoms are severe and non-pharmacological interventions are not enough.

SSRIs are both safe and efficacious. Common SSRIs include fluvoxamine, fluoxetine, and sertraline.

Tricyclics are indicated with careful monitoring of cardiac functioning.

Benzodiazepines should be used only when a rapid reduction of symptoms is needed.

Selective Mutism

Selective mutism is the failure to speak where there is an expectation for speaking in social situations.

The individual chooses to communicate via non-verbal means such as pointing, gesturing, nodding, whispering, writing, etc. The inability to speak is not due to lack of knowledge of the language.

Collaboration between the speech-language pathologist, and other health care providers (clinical psychologist, social worker, psychiatrist), classroom teacher and child's family, is necessary for assessment and treatment planning. Other conditions such as speech and language delays must also be considered.

Selective mutism may coexist with social anxiety, separation anxiety, and OCD.

Individuals with timid personality traits might be prone to develop selective mutism.

Cultural Issues

New immigrants to a country where a different language is spoken may refuse to speak the new language because of lack of knowledge of the language.

Higher prevalence rates of selective mutism have been noted in immigrant children and in language minority children when compared to non-immigrant children.

Sometimes it is a way of self-protection during an experience of intense anxiety.

Accurate diagnosis of selective mutism can be difficult due to the initial nonverbal stage (i.e., "silent period") common to second language learners; they appear quiet as they focus on listening and comprehension [24].

Consider school stressors, when treating an English Learner with selective mutism, including lack of class support for learning a second language, negative views of the child's culture or language, limited communication between parent/caregiver and school.

Bilingual children with true selective mutism present mutism in both languages in several unfamiliar settings and for significant periods of time. Interviewing parents and caregivers to determine if the child speaks in his/her language successfully outside of the home environment is important information [25].

Consider determining educational eligibility within the public school setting under the Individuals with Disability Education Act of 2004 (IDEA 2004), in the categories of other health impairment, speech-language impairment, or emotional disturbance/disability.

If the child does not qualify for an individualized educational program (IEP) then consider whether a 504 Plan would meet some of the child's needs in the classroom.

Epidemiology

Selective mutism is a rare disorder.

Prevalence estimates range between 0.03% and 1% depending on the setting (e.g., clinic vs school vs general population). It does not seem to vary by sex or race/ethnicity.

It is more common in 3–6-year-old.

The disorder is more likely to manifest in young children than in adolescents and adults.

Diagnosis

Screening individuals who present with language and communication difficulties to determine need for further assessment and/or referral for other services needed include:

Collateral information from parent/caregiver to describe home behaviors, routines, eating and sleeping patterns, teacher report from classroom observations.

Hearing screening to rule out hearing loss.

Psychologist evaluation to guide IQ testing for learning difficulties and other mental conditions such as autism spectrum disorder and schizophrenia.

Treatment

The goal of treatment is to decrease anxiety and to increase verbal communication.

Early and consistent behavioral interventions at home and school are key to remediation. Adults and other children are not to speak for the affected individual.

Providing predictability and control for the child will decrease anxiety.

A multidisciplinary team approach consistent of pediatrician, psychologist, language pathologist, teacher, counselor, and family's report is necessary to evaluate speech, language, and communication domains in children with selective mutism.

Affected children may require personalized treatment sessions to establish rapport and to practice relaxation techniques in a safe and comfortable setting.

Behavioral strategies to explore include: graded exposure to anxiety provoking situations, systematic desensitization along with relaxation techniques, stimulus fading and contingency management, giving positive reinforcement upon verbalization.

Adolescents and Adults.

Treatment of adolescents and adults with selective mutism can prove difficult.

Cognitive behavioral therapy and motivational interviewing could be beneficial.

Consider anti-anxiety medications to address social anxiety if appropriate.

Specific Phobia

It is a persistent fear and anxiety of a specific object or situation, which may be expressed as fear of losing control, or getting harmed, or having dizziness or fainting spells.

Typical phobic objects or situations include heights, storms, enclosed places, animals (e.g., snakes, bugs), witnessing blood, driving on bridges and through tunnels among some examples.

Cultural Issues

Hispanics/Latinos report significantly lower rates of specific phobias than non-Latino whites.

Epidemiology

World Mental Health Survey [26] suggests specific phobia is one of the most common disorders in the general population with a lifetime prevalence ranging from 7.7% to 9% in the United States; in South American countries it is 2%–4%.

Prevalence rates are approximately 5% in children and 16% in 13–17-year-old.

Females are more frequently affected than males.

The median age of onset is 8 years of age.

Specific phobia is rarely seen in medical-clinical settings in the absence of other psychopathology.

Treatment

Cognitive Behavioral Therapy (CBT) with cognitive restructuring, and gradual desensitization through increased exposure to the stimulus can be effective treatment and relaxation techniques.

Exposure response prevention is the main form of treatment [27, 28].

Selective serotonin reuptake inhibitors and benzodiazepines are effective medications to consider.

Social Anxiety Disorder (Social Phobia)

Social anxiety is a chronic condition with excessive fears in social situations when exposed to scrutiny by others. It might last a lifetime, rates of complete remission are rare.

Typical social situations include fear of public speaking or performing in public, fear of being criticized for poor school or work performance.

Individuals are sensitive to criticism, and they have low self-esteem and poor social skills.

Social anxiety may coexist with avoidant personality, making it difficult to distinguish between the two.

Cultural Issues

Immigrant status has lower rates of social anxiety disorder in both Latino and non-Latino white groups. Individuals living in urban or populated regions have lower risk.

Lower risk among minorities is more pronounced at lower levels of education.

Epidemiology

The 12-month prevalence estimate of social anxiety for the United States ranges from 2.8% to 7% depending on the study but is still considerably higher than in most non-European countries.

Lower 12-month prevalence is seen in other parts of the world, around 0.5%–2%. In Europe it is 2.3%. In Mexico, it was 1.7% [29]. Mean age at onset is about 15 years. Social anxiety is unlikely to begin after age 30.

The lower risk among Latinos, relative to non-Latinos whites in the United States, was found only among the young cohort (age < 43 years) [30].

Treatment

The most effective treatment [31] is the combination of Selective Serotonin Reuptake Inhibitors (SSRIs) with Cognitive Behavioral Therapy (CBT).

SSRIs including sertraline (Zoloft), paroxetine (Paxil), and fluvoxamine (Luvox) are first-line medications for treatment of social anxiety (paroxetine has more evidence than any other SSRI, but other SSRIs are similar including fluvoxamine, sertraline, fluoxetine, and citalopram).

SSRIs are generally well tolerated, easy to dose, and relatively safe in overdose and their ability to treat co-morbid conditions.

Other medications to consider are: SNR: Venlafaxine (Effexor), Beta Blockers: Propranolol (Inderal), Buspirone.

Benzodiazepines: Clonazepam has the largest effect size of all medications. There are concerns of excessive sedation, potential with withdrawal and dependence and risks or relapse after discontinuation [32].

Augmenters/alternatives in treatment consider: gabapentin, pregabalin.

Psychosocial treatment interventions.

Cognitive Behavioral Therapy (CBT) including: cognitive restructuring, exposure response prevention, relaxation techniques.

Social Rehabilitation: social skills training (individual or group format), communication and assertiveness training, role playing, joining clubs.

Panic Disorder

Panic disorder has panic attacks several times a week. Panic attacks can last minutes or hours. Symptoms can occur from a calm state or an anxious state. It is a chronic condition with waxing and waning course and intensity.

Different medical disorders, psychiatric disorders, substance abuse disorders, and medications can induce panic attacks. Panic attacks are a symptom and do not confer a diagnosis of panic disorder necessarily.

Patients with panic disorders are first seen by primary care or in Emergency departments with frequent visits due to physical and emotional symptoms.

Panic disorder is frequently underrecognized.

Panic Attack Specifier

Panic attacks could occur with any anxiety disorder as well as other mental disorders (e.g., depressive disorders, post-traumatic stress disorder, substance use disorders) and some medical conditions (e.g., cardiac, respiratory, vestibular, gastrointestinal).

It may be expressed with autonomic hyperarousal from different body systems including:

Cardiac: chest pain, palpitations, chills/hot flushes, fast heart rate, sweating. Consider a correlation between panic disorder and mitral valve prolapse, Echocardiogram is diagnostic.

Neurological: shaking, dizziness, numbness, spells. There is some correlation with seizure disorders. EEG is diagnostic.

Respiratory: choking sensations, shortness of breath, asthmatic crisis.

Gastrointestinal: nausea, abdominal discomfort, diarrhea.

Others: Derealization, depersonalization, fear of losing control or dying.

Cultural Issues

Panic attacks might be an expression of cultural distress in Hispanics.

Cultural syndromes among Latinos including Ataque de Nervios are associated with panic disorder often.

Some clinical presentations of "Ataque de Nervios" fulfill criteria for conditions other than panic attacks including dissociated disorder, conversion disorder, or anxiety NOS.

Epidemiology

The lifetime prevalence of panic disorder in the United States is 4.7% to 5.1%.

It was 0.6% in Mexico.

Two peaks of onset are between ages 17 and 35 years. More common in females.

Onset after age 40 suggests an underlying medical condition or depression.

In the United States, significantly lower rates of panic disorder are reported among Latinos, Caribbean blacks when compared to non-Latino whites.

Rates among Puerto Ricans are higher 4.9% than in other Latino subgroups 2.1%–3.2%.

The causes of these cross-cultural differences and differences in prevalence remain unclear.

In the survey samples, more respondents received an anxiety disorder NOS diagnosis than all specified anxiety.

Treatment

Recommended treatment for panic disorder involves the use of antidepressants SSRIs, SNRI, benzodiazepines, TCAs, augmentation strategies, combined with panic-focused CBT.

SSRIs: Fluoxetine (Prozac), Sertraline (Zoloft), and Paroxetine (Paxil) are FDA approved for the treatment of panic disorder. Consider combination or augmentation strategies (e.g., an SSRI + a TCA, Trazodone, Buspirone, a benzodiazepine, or an atypical antipsychotic medication) for optimization as needed.

SNRI: Venlafaxine XR (FDA approved).

Benzodiazepines are the most common treatment prescribed.

Alprazolam (Xanax) and Clonazepam (Klonopin) are FDA approved for treatment of panic disorder.

The potential for abuse should always be considered prior to initiating treatment with a benzodiazepine.

Combining a benzodiazepine and an SSRI can be especially effective.

Tricyclics (TCAs), imipramine, and clomipramine are effective for panic disorder (Not FDA approved for this indication). Side effects may limit their usefulness.

Panic-Focused Cognitive Behavioral Therapy (CBT) is an effective treatment alone or in combination with anti-panic medications for panic disorder. Sessions focused on cognitive restructuring, breathing, and bodily sensation awareness.

Agoraphobia

It is excessive anxiety of being in places where escape or help might be difficult including:

Public transportation, open and enclosed spaces, crowded gatherings, or being outside alone.

Agoraphobic fears may decrease in the presence of a trusted individual.

Anxiety is out of proportion to the actual danger.

Agoraphobia is diagnosed irrespective of the presence of panic disorder.

Cultural Issues

The cross-racial/ethnic variability of agoraphobia without panic disorder in the United States has received little attention.

Prevalence rates of agoraphobia do not appear to vary systematically across cultural/racial groups.

Epidemiology

The 12-month prevalence of agoraphobia is approximately 1.7% of adolescents and adults in the United States. It was 0.7% in Mexico [1].

Latinos and non-Latino whites report similar lifetime prevalence, but Puerto Ricans endorse a significant higher rate 6% than other Latino groups 2.1%–3.2% [33]. Females are twice as likely as males to experience agoraphobia.

Agoraphobia may occur in childhood, but incidence peaks in late adolescence and early adulthood. For older individuals, it is 0.4%.

Generalized Anxiety Disorder (GAD)

Anxiety is intense and relates to life matters (e.g., health, family, work, social, financial).

It consists of psychological and physical symptoms including: constant worries, restlessness, irritability, lack of concentration, fatigue, muscle tension: body, neck, back pains, headaches, sleep difficulties.

If anxiety is attributed to an identifiable cause with short duration, consider an adjustment disorder with anxiety.

If anxiety is due to a substance or medication, consider a substance/medication-induced anxiety disorder.

If panic attacks or anxiety is a consequence of a medical condition, consider anxiety disorder due to another medical condition.

Generalized anxiety disorder with/without panic attacks are highly co-morbid (up to 50%) with many medical, mood disorders, alcohol use, and other anxiety disorder.

Cultural Issues

The expression of Generalized Anxiety Disorder (GAD) has cultural variations.

Hispanics might tend to exhibit more somatic symptoms than cognitive symptoms.

Epidemiology

In the United States, the lifetime prevalence is estimated to be about 5%, some discrepancy in community-based rate studies 2.1%–2.9%. It was 0.4% in Mexico [29]. GAD is seen in women twice as frequently as in men.

Age of onset is variable.

Some individuals reported in childhood and before age 20.

The Epidemiologic Catchment Area (ECA) study (1980–1985) found that lifetime rates of DSM-III GAD varied with non-Latino whites reporting twice the prevalence than US-born Mexican Americans and over four times the prevalence reported by Mexican immigrants 6.9% vs 3.4%. (do we need a reference here?)

This discrepancy was attributed to Hispanics participants inability to understand key instrument items, lack of semantic equivalence in instrument translation rather than symptom variability.

The most common somatic subscale of the Beck Anxiety Inventory (BAI) endorsed by Hispanics was dizziness and indigestion. The US non-Hispanic participants scored higher on the psychological subscale "scared" and "nervous."

A cross-national comparison study of Mexican youth in Mexico reported more anxiety-related somatic symptoms than Euro-American youth [34].

Differences in reports of somatic versus psychological symptoms associated with GAD have been found in the same individuals, depending on the language used to report their symptoms.

The finding that bilingual Latinos tend to use somatic expressions to describe their GAD symptoms in Spanish and psychological ones to describe the same condition in English suggests that Spanish-monolingual Latinos with GAD may be less likely to be identified, if largely psychological criteria are adopted for the disorder [35].

Psychological Testing for Generalized Anxiety Disorder and Panic Disorder.

Measure scales are used for screening and monitoring of symptoms severity and cannot replace a clinical assessment and diagnoses. The cross-cultural validity of instruments assessing anxiety/worry is limited and has produced mixed results.

The most widely used scales are:

Beck Anxiety Inventory is a 21 question multiple choice self-report inventory to measure severity of anxiety in children and adults. It has been criticized for its predominant focus on physical symptoms of anxiety. It may not function well for social anxiety or OCD which have stronger cognitive or behavioral component. There is a Spanish version [36].

Hamilton Anxiety rating scale (Ham-A) was developed to measure the severity of anxiety symptoms. The scale consists of 14 items, each defined by a series of symptoms, and measures both psychic anxiety (mental agitation and psychological distress) and somatic anxiety (physical complaints related to anxiety). It remains widely used as an outcome measure in clinical trials.

It has been translated into Spanish.

Self-rating scales:

GAD-7 is an easy to use self-administered patient questionnaire with seven items on 4-point scale. Using the threshold score of 10, the GAD-7 has a sensitivity of 89% and a specificity of 82% for GAD, it is moderately good at screening panic disorder, social anxiety, and PTSD.

It has a Spanish version.

GAD-2 item scale is also available.

Panic attack scale is a 7-question scale to measure severity of panic attacks using 0 to 4 for scoring. Any score over 9 is considered important. It has a Spanish version but few data exist on this instrument in non-white American samples [37].

Treatment

SSRIs: Escitalopram (Lexapro) and Paroxetine (Paroxetine, Pexeva) are FDA approved for GAD. Consider initiating treatment at a lower than therapeutic dose with gradual increase to avoid initial worsening anxiety.

SNRI: Venlafaxine (Effexor), Duloxetine (Cymbalta)(FDA approved for GAD).

Buspirone: (Buspar): FDA approved, it is a serotonergic agonist agent effective for treatment of GAD and for the maintenance treatment of GAD, no addiction potential.

Benzodiazepines: Alprazolam (Xanax), Clonazepam (Klonopin) (FDA approved) clinically indicated for anxiety symptoms and insomnia, augmentation of antidepressants, mood stabilizers, and antipsychotic medications, alcohol and hypnotic withdrawals, seizures, muscle strain. Rapid onset of action, favorable side effect profile. Because of potential for dependence, recommendations are for short-term rather than long-term use and may cause disinhibition, memory deficits, and dizziness.

Most agents are cross tolerant with each other.

Other Medications TCAs and other heterocyclics: are not FDA approved for the treatment of GAD. However, imipramine (Tofranil), doxepin (Sinequan), and trazodone (Desyrel) can be effective agents.

Hydroxyzine (Vistaril, Atarax): Anti-histaminergic medications can decrease anxiety, nausea, vomiting, allergies, skin rash, hives, and itching.

Psychotherapeutic approaches:

Include insight-oriented or supportive psychotherapy, interpersonal therapy, cognitive behavioral therapy (CBT) with cognitive restructuring, exposure response prevention, relaxation training, group therapy, yoga, meditation, and lifestyle changes (e.g., reduction of stressful environment, improved sleep hygiene, exercise).

Substance/Medication-Induced Anxiety Disorder

Panic attacks/Anxiety are seen often in the presentation.

The panic attacks developed after a substance intoxication or withdrawal or after exposure to a substance/medication.

Epidemiology

Prevalence of substance/medication-induced anxiety disorder is not clearly understood.

It may be rare, 0.002%. No data found for Hispanics.

Anxiety Disorder Due to another Medical Condition

Panic attacks/Anxiety are seen often in the presentation.

Panic attacks are the direct pathophysiological consequence of another medical condition.

Epidemiology

Prevalence of anxiety disorder due to another medical condition is not clearly understood. Patients with asthma, hypertension, ulcers, and arthritis are common medical conditions with high risk of anxiety disorders.

Other Specified Anxiety Disorder

It applies to an anxiety presentation that does not meet the full criteria for any of the disorders in the anxiety disorders diagnostic class. It is also used when the clinician chooses to communicate specific reasons for not meeting criteria (e.g., Ataque de Nervios).

Unspecified Anxiety Disorder

It applies to an anxiety disorder that does not meet the full criteria for any of the disorders in the diagnostic class. The clinician chooses Not to specify the reason for not meeting criteria. (e.g., in emergency room setting, lack of information).

In past studies, individuals from minority groups, including Hispanics, did not meet full criteria for specified anxiety disorders and therefore had lower prevalence rates when compared to white Americans.

It is possible that the WMH-CIDI (World Mental Health Survey Initiative version of the World Health Organization Composite International Interview) falls short across language and cultural differences to capture the anxiety in the minorities [38].

Other Relevant Anxiety Topics to Consider

A.Recurrent thoughts and repetitive behaviors, as in Obsessive Compulsive Disorder (OCD).

Cross-cultural homogeneity data with DSM criteria found 1% in the United States, 1.8% in Puerto Rico. Data on cross-cultural assessment of OCD severity and impairment are very scarce. OCD epidemiology indicates similarity in gender distribution, age of onset, and comorbidity. OCD is common in different cultural settings in male children and adolescents and in female adults. Late adolescence is a period of increased vulnerability, and individuals with OCD may also have mood and anxiety disorders.

B.Re-experiencing Trauma, as in Post-Traumatic Stress Disorder (PTSD). Many Central Americans immigrants are at high risk for PTSD for war-related trauma and terror that preceded their immigration to the United States and may make adjustment to their new home more difficult. In a national study of Vietnam veterans [39], Kulka et al. found that Hispanics were at higher risk for PTSD than their white counterparts. Ortega and Rosenheck [15] studied Puerto Ricans veterans in particular and found to have higher probability of experiencing PTSD than others with similar levels of war zone stress.

C.Preoccupation with appearance, as in Body Dysmorphic Disorder. Data on cross-cultural.

Assessment of Body Dysmorphic Disorder is scarce. Adolescent girls were most dissatisfied with their bodies than adolescent boys. The interaction between gender and ethnicity was not significant [40].

D.Difficulty discarding or parting with possessions, as in Hoarding Disorder. The majority of studies have been conducted in samples with white/Caucasian individuals. Data on Hispanics is scarce [41].

E.Hair Pulling, as in Trichotillomania. A review of the literature suggested that research on Latinos/Hispanics is limited to case studies and first person accounts [42].

F. Skin Picking, as in Excoriation Disorder. A Pubmed review of the literature showed no reports on Hispanics.

The limited literature is reflective of the general state of research involving minorities and Anxiety Disorders, more research needs to be done to test guidelines, standards, or models to develop sensitive interventions for Hispanic/Latinos and other minorities.

References

1. Lewis-Fernández R, Hinton DE, Laria AJ, Patterson EH, Hofmann SG, Craske MG, et al. Culture and the anxiety disorders: recommendations for DSM-V. Depress Anxiety. 2010;27(2):212–29.
2. US Census Bureau [Internet]. [cited 2020 Jun 22]. Available from: https://www.census.gov/topics/population/hispanic-origin.html
3. O'Connor A, Batalova J, Bolter J. Central American immigrants in the United States [internet]. Migration Policy Institute. 2019; Available from: https://www.migrationpolicy.org/article/central-american-immigrants-united-states
4. Educational Attainment 2000 [Internet]. [cited 2020 Jun 22]. Available from: https://www.census.gov/prod/2003pubs/c2kbr-24.pdf
5. Key facts about US immigration policies and proposed changes [Internet]. [cited 2020 Jun 24]. Available from: https://www.pewresearch.org/fact-tank/2019/05/17/key-facts-about-u-s-immigration-policies-and-proposed-changes/
6. Collins MMD and SR. Millions More Latino Adults Are Insured Under the Affordable Care Act [Internet]. [cited 2020 Jun 29]. Available from: https://www.commonwealthfund.org/blog/2017/millions-more-latino-adults-are-insured-under-affordable-care-act
7. US Census Bureau [Internet]. [cited 2020 Jun 25]. Available from: https://www.census.gov/library/stories/2019/02/hispanic-poverty-rate-hit-an-all-time-low-in-2017.html
8. Canino G, Shrout PE, NeMoyer A, Vila D, Santiago KM, Garcia P, Amarilis Quiñones VC& MA. A comparison of the prevalence of psychiatric disorders in Puerto Rico with the United States and the Puerto Rican population of the United States. Soc Psychiatry Psychiatr Epidemiol. 2019;54:369–78.
9. Pew Research Center. Religion and Public Life [Internet]. [cited 2020 Jun 25]. Available from: https://www.pewforum.org/2019/10/17/in-u-s-decline-of-christianity-continues-at-rapid-pace/
10. Sabogal F, Marín G, Otero-Sabogal R, Marín BVO, Perez-Stable EJ. Hispanic Familism and acculturation: what changes and what Doesn't? Hisp J Behav Sci. 1987;9:397–412.
11. Randy A, Sansone MD, Lori A, Sansone M. Psychiatric disorders. Psychiatry (Edgmont). 2010;7(12):16–9.
12. Peter J. Guarnaccia PhD IMB& HA. Mental health in the Hispanic immigrant community. J Immigr Refug Serv. 2005;
13. Glover SH, Pumariega AJ, Holzer CE, Wise BK, Rodriguez M. Anxiety symptomatology in Mexican-American adolescents. J Child Fam Stud [Internet]. 1999;8(1):47–57. Available from: https://www.academia.edu/9029877/Anxiety_Symptomatology_in_Mexican-American_Adolescents
14. Escobar JI, Burnam MA, Karno M, Forsythe A, Golding JM. Somatization in the community. Arch Gen Psychiatry. 1987;44(8):713–8.
15. Ortega AN, Rosenheck R, Alegría M, Desai RA. Acculturation and the lifetime risk of psychiatric and substance use disorders among Hispanics. J Nerv Ment Dis. 2000;188(11):728–35.
16. Malgady RG, Rogler LH, Costantino G. Ethnocultural and linguistic bias in mental health evaluation of Hispanics. Am Psychol. 1987;42(3):228–34.
17. Malgady RG, Costantino G. Symptom severity in bilingual hispanics as a function of clinician ethnicity and language of interview. Psychol Assess. 1998;10(2):120–7.
18. Guarnaccia PJ, Rubio-Stipec M, Canino G. Ataques de nervios in the Puerto Rican diagnostic interview schedule: the impact of cultural categories on psychiatric epidemiology. Cult Med Psychiatry. 1989;13(3):275–95.
19. Salgado De Snyder VN, Diaz-Perez MDJ, Ojeda VD. The prevalence of nervios and associated symptomatology among inhabitants of Mexican rural communities. Cult Med Psychiatry. 2000;24(4):453–70.
20. Vega WA, Khoury EL, Zimmerman RS, Gil AG, Warheit GJ. Cultural conflicts and problem behaviors of Latino adolescents in home and school environments. J Community Psychol. 1995;23(2):167–79.

21. Suårez-Orozco C, Todorova ILG, Louie J. Making up for lost time: the experience of separation and reunification among immigrant families. Fam Process. 2002;41(4):625–43.
22. Ginsburg GS, Silverman WK. Phobic and anxiety disorders in Hispanic and Caucasian youth. J Anxiety Disord. 1996;10(6):517–28.
23. Separation Anxiety Facts and Statistics [Internet]. [cited 2020 Jun 29]. Available from: https://www.therecoveryvillage.com/mental-health/separation-anxiety/related/separation-anxiety-statistics/
24. Elizur Y, Perednik R. Prevalence and description of selective mutism in immigrant and native families: a controlled study. J Am Acad Child Adolesc Psychiatry. 2003;42(12):1451–9.
25. Cohan SL, Chavira DA, Stein MB. Practitioner review: psychosocial interventions for children with selective mutism: a critical evaluation of the literature from 1990-2005. J Child Psychol Psychiatry Allied Discip. 2006;47(11):1085–97.
26. Wardenaar KJ, Lim CCW, Al-Hamzawi AO, Alonso J, Andrade LH, Benjet C, et al. The cross-national epidemiology of specific phobia in the world mental health surveys. Psychol Med. 2017;47(10):1744–60.
27. Ameican Psychiatric Association. Diagnostic and statistical manual of mental disorders Fifth Edition: DSM-5. 5th ed.
28. Sadock B, Sadock VA. Kaplan & Sadock's concise textbook of clinical psychiatry. Wolters Kluwer: Philadelphia; 2017.
29. Medina-Mora ME, Borges G, Lara C, Benjet C, Blanco J, Fleiz C, Villatoro J, Estela Rojas JZ. Prevalence, service use, and demographic correlates of 12-month DSM-IV psychiatric disorders in Mexico: results from the Mexican National Comorbidity Survey. Psychol Med. 2006;35(12):1773–83.
30. Breslau J, Aguilar-Gaxiola S, Kendler KS, Su M, Williams D, Kessler RC. Specifying race-ethnic differences in risk for psychiatric disorder in a US National Sample. Psychol Med. 36(1):57–68.
31. Blanco C, Schneier FR, Schmidt A, Blanco-Jerez CR, Marshall RD, Sánchez-Lacay A, et al. Pharmacological treatment of social anxiety disorder: a meta-analysis. Depress Anxiety. 2003;18(1):29–40.
32. Connor KM, Davidson JRT, Potts NLS, Tupler LA, Miner CM, Malik ML, et al. Discontinuation of clonazepam in the treatment of social phobia. J Clin Psychopharmacol. 1998;18(5):373–8.
33. Alegría M, Canino G, Shrout PE, Woo M, Duan N, Vila D, et al. Prevalence of mental illness in immigrant and non-immigrant U.S. Latino groups. Am J Psychiatry. 2008;165(3):359–69.
34. Varela RE, Sanchez-Sosa JJ, Biggs BK, Luis TM. Anxiety symptoms and fears in hispanic and European American children: cross-cultural measurement equivalence. J Psychopathol Behav Assess. 2008;30(2):162. https://doi.org/10.1007/s10862-007-9056-y.
35. Hirai M, Stanley MA, Novy DM. Generalized anxiety disorder in hispanics: symptom characteristics and prediction of severity. J Psychopathol Behav Assess. 2006;28(1):49–56.
36. Sanz J, Navarro ME. Propiedades psicométricas de una versión española del Inventario de Ansiedad de Beck (BAI) en estudiantes universitarios. Ansiedad estrés. 2003;9(1):59–84.
37. Fuste G, Gil MÁ, López-Solà C, Rosado S, Bonillo A, Pailhez G, et al. Psychometric properties of the Spanish version of the panic disorder severity scale. Span J Psychol. 2018;21:E5.
38. Asnaani A, Richey JA, Dimaite R, Hinton DE, Hofmann SG. A cross-ethnic comparison of lifetime prevalence rates of anxiety disorders. J Nerv Ment Dis. 2010;198(8):551–5.
39. Kulka RA, Schlenger WE, Fairbank JA, Hough RL, Jordan BK, Marmar CR, et al. Trauma and the Vietnam war generation: report of findings from the national Vietnam veterans readjustment study. Trauma Vietnam War Gener Rep Find from Natl Vietnam Veterans Readjustment Study. 2014;18:1–322.
40. Mayville S, Katz RC, Gipson MT, Cabral K. Assessing the prevalence of body dysmorphic disorder in an ethnically diverse group of adolescents. J Child Fam Stud. 1999;8(3):357–62.
41. Fernández de la Cruz LE, Nordsletten A, Mataix-Cols D. Ethnocultural aspects of hoarding disorder. Curr Psychiatr Rev. 2016;12(2):115–23.
42. Neal-Barnett A, Flessner C, Franklin ME, Woods DW, Keuthen NJ, Stein DJ. Ethnic differences in trichotillomania: phenomenology, interference, impairment, and treatment efficacy. J Anxiety Disord. 2010;24(6):553–8.

The Role of Acculturation in the Mental Health of Hispanics

7

Renato D. Alarcón

Introduction

Acculturation is a term of fluid use in a variety of social sciences, historical and even philosophical topics [1, 2]. Although preferentially incorporated in the literature related to migration phenomena and their resulting changes, adjustments and adaptations across time, location and context [3, 4], it is also the subject of journalistic inquiries, methodological documents, social policies, political debates, and even fiction masterpieces. From a predominantly psychiatric perspective, acculturation occupies significant spaces in diagnostic and therapeutic areas: it plays a pathogenic role as a precipitating factor of a number of clinical conditions [5], and can be the source of therapy-oriented approaches for a variety of emotional disorders in members of different ethnic or cultural groups.

In spite of its very long history as a unique process of human interactions, the systematic and consistent study of acculturation may have started only in the first decades of the twentieth century when social scientists fixed their research interests in the complex process of world migrations and their many modalities, causal factors, evolvement, and outcomes. Culture, traditionally defined as a repository of experiences, traditions, habits, religion, values, and identities of human groups or communities [6, 7], entails also as a dynamic process, a changeable scenario resulting from the growing interactions between migrant groups and host collectivities [8]. Such unavoidable interactions take place in a variety of circumstances, dressed up by a variety of emotions: curiosity, predisposing aversion, hatred, fear, resignation, wish to help or compassion, among the members of the host society; and expectations, hope, illusions, uncertainty, fear, defensiveness, or anxiety in the migrant population. The concept, practice, and study of acculturation, thus

R. D. Alarcón (✉)
Emeritus Professor of Psychiatry, Mayo Clinic School of Medicine, Rochester, MN, USA

Honorio Delgado Chair, Universidad Peruana Cayetano Heredia, Lima, Peru
e-mail: Alarcon.Renato@mayo.edu; Renato.Alarcon@upch.pe

R. Castilla-Puentes, T. Falcone (eds.), *Mental Health for Hispanic Communities*,
https://doi.org/10.1007/978-3-031-13195-0_7

generated, defined it as the uniquely contextualized process of experiential exchanges between the two protagonist groups of the migratory encounter, the resulting set of habits, beliefs, customs and everyday life practices in migrants and hosts [9].

There are many approaches to the study of the acculturative phenomena, depending on a variety of methodologies and perspectives covering different magnitudes of change [4, 5, 10]. The first criterion entails a causal approach, i.e., the reasons compelling people to take the route of migration: Was it voluntary or forced? If the former, be that determined by economic, occupational, deliberate factors and more or less planned organization, an open, amenable disposition to face the acculturative process and its specific experiences can be expected. If forced by social, financial, political reasons, or even by natural disasters, migration can become a profoundly traumatic event with refugee or displaced crowds unwillingly compelled to abandon familiar places and enter into unknown, even hostile scenarios [11, 12]: acculturation would then be an unexpected, imposed, multiplied traumatic process. An additional component has to do with whether the migration is internal (i.e., within a known territory such as the same country or region) or external, implying the crossing of international borders and the confrontation with more complex, testing, and demanding circumstances.

Among the many types or cases of migratory phenomena across world regions or continents, the movement of millions of Hispanic or Latin American *(Latino)* groups towards North America (mostly the United States) has a number of unique characteristics [13]. To cite a few of them, it can be said that most of the migrants come from Central American countries and Mexico, most are of modest socio-economic population segments, take on the route pushed by labor and work needs, and use either plain walking or crowded transportation means (i.e., buses, trucks, or trains) for their displacement to the Mexico–US border. At that point, they start probably the most complicated phase of the process, choosing either an uncertain chain of bureaucratic procedures, or the desperate and costly illegal entry led by demanding, abusive, unscrupulous "guides," properly called *coyotes* [14]. These immigrants are mostly adult males, but thousands of women, adolescents, children, and even senior people also make this dramatic journey.

This chapter will briefly examine different aspects of the acculturation of Hispanic immigrants in the USA, described both as a socio-cultural process and as a specific human experience [10, 15, 16]. It will mostly focus on the health (physical and, mostly, mental or emotional) implications of acculturation, centering on the acculturative stress as both a critical component of the adaptational process, and a clinical condition, eventually recognized by classification manuals and other diagnostic resources. It will finish outlining management steps when necessary, and elaborating didactic conclusions and recommendations related to the various perspectives on the topic.

The Process and the Experience of Acculturation

Individual and collective changes induced by acculturation have been systematically studied through a variety of models and across several domains. The process generates changes in attitudes, behaviors, values, and cultural identity. Many authors mention acculturative motivations, availability of resources, and modeling factors as varied as personality dispositions, vocational interests, food preferences, peer relations, and media usage patterns [17–19]. In turn, these acculturation models vary in function of uni- or pluri-dimensional modeling factors. Perhaps the most consistently accepted approach to the study of this issue is Berry's orthogonal, two-factor model of acculturation [20, 21] determined by the ethnic-heritage, and the dominant-culture mainstream dimensions whose dynamic, reciprocal interactions produce four distinctive, grade-based acculturative states: integration, assimilation, separation, and marginalization. Like in many other fields of scientific inquiry, this approach is not unanimously accepted, as purely socio-demographic and socio-economic factors seem to play a larger role than acculturation itself on the mental health implications of migration [22]. Furthermore, moderating factors such as cultural intelligence and perceived diversity climate of the host entities seem to be powerful pivots of a positive international acculturation [23].

The main components of acculturation as a *process* are, of course, the protagonists of the encounter: members of the migrant group and the host community or society. As important as they are, however, a crucial component is the set of circumstances that resulted in the migration itself, that is, a series of factors that evolve, silently or noisily, as triggers of a more or less predictable type of reception. For instance, if the migration was related to political situations (civil wars, dictatorships, persecutions, formal or violent exile, etc.), the reaction of the host society may include either sympathy, compassion and open support or rejection, distrust or different levels of uncertainty [19, 24, 25]. On the other hand, if migration responds to economic needs, joblessness in the country or region of origin, or pervasive levels of poverty and subsequent violence, the migrants will do it out of desperation but still embracing levels of hope, while the hosts may feel this as an opportunity of increasing its workforce at a relatively low cost, or as an inappropriate attempt to take over existing jobs [26, 27]. Whether the migration process takes place within or outside legal frames of reference, adds an important conditioning factor to the process.

The setting in which acculturation occurs is also a critical component. It has to do with the part of the host community where the immigrant arrives or decides to settle down. Massive migrations like that of Hispanics into the United States, usually lead to more or less deprived urban or peripheral areas in large cities [28] to the point that they seem unable to establish direct contact with members of the host communities: however, the fact is that the acculturation process starts almost immediately, regardless of whether the protagonists are or are not aware of its occurrence [29, 30]. Even though living in areas with high concentration of Latinos seems to make acculturation advance more slowly [28, 31], newly arrived immigrants are constantly reminded of the fact that they are in a different milieu, facing a different reality, subjected to new rules that they may, at first, attempt to deny or minimize, compelled to familiarize themselves with sets of psychosocial boundaries (i.e., language, food or communication styles, gender relationships, family life, authority principles, educational and religious practices, etc.) that in the end will generate a different behavioral repertoire [32, 33]. Acculturation as a process, however, may be more a portrait, an academic description, even an animated narrative.

As an *experience*, on the other hand, acculturation is, in many cases, an agitated, dynamic, at times chaotic succession of events, a personal story of intense learning and confronting circumstances and situations whose outcome remains uncertain for long periods. The constantly changing contexts (neighborhood, school, workplaces, churches, streets, transportation) challenge the adaptive capacities of migrant individuals and groups in a sequence that may be extremely disorganizing [17–19, 26, 28, 34, 35]. The environment into which the immigrant has arrived makes him/her aware of his otherness [36, 37], of his precarious standing of anonymity that, paradoxically, is not going to make him/her less visible, quite the contrary. And the problem becomes even more complicated because the immigrant is not or, many times, does not want to be aware of this collective ambiguity.

Acculturation is, therefore, a process that defies polite academic descriptions because it is constituted by intense experiences and variations. The most sophisticated research describes different outcomes, mostly qualified by the degree in which the newcomer adopts, absorbs, tolerates or rejects the social and cultural rules of the host society, against the background of individual personality, mood, cognitive, and other psychobiological traits [30, 31, 38–40]. Many authors consider the extremes of this process (i.e., total, almost unconditional acceptance vs. total, almost absolute rejection) as undesirable. The former, because it may reflect total subjugation or subordination to the "new" norms and the forging of a new form of renegade renouncing to history and legacies; the latter, because it would show unwillingness to reasonable adaptations, an unrealistic denial of the practicality of changing to advance without abandoning the telluric essences of one's own self.

Acculturation in Hispanic Communities

Perhaps more than any other ethno-cultural group among migrant communities, Hispanics (or *Latinx*) coming into the United States offer the best examples of the good and bad sides of the acculturative phenomenon. Currently being the largest minority population in the country and having made unquestionable progress in a variety of areas, it is clear however that a lot remains to be done [41, 42]. The majority of Hispanic immigrants are from Mexico and Central America. A large proportion continues to be immersed in deep levels of poverty, unstable occupational and labor positions, deprived living conditions and risks of discrimination, stigmatization and neglect [43, 44]. A good percentage of the Hispanic population in the USA is made out of immigrants mostly devoted to rural, agricultural or moving work locations with predictable family instability, financial limitations, and scarce opportunities of physical, social, and emotional well-being [27, 29, 33, 45]. Its political impact is still far lower than the other largest minority group—African-Americans—and it is not yet possible to speak of strong or consistent efforts to coordinate and solidify common objectives.

A lot has been studied and written about the strengths and weaknesses of the Hispanic population and communities in the USA. *Familism* continues to be a well-preserved value among Latinos, together with its solid connections to respect for the elderly, the mother's figure (and its resulting *Marianism*), and strong religious (mostly Catholic) devotions and practices [25, 46, 47]. On the other hand, economic factors generate many times family separations, particularly at the start of the migration process, as fathers and adolescents come first leaving the rest of the family awaiting and hoping. The same circumstances may contribute to nutritional deficits and lack of healthy physical practices, disagreements, criticisms, and eventual breaks of domestic violence [43, 45, 48]. Interestingly, there are differences between Hispanic immigrant groups (South Americans, Central Americans, Caribbeans), i.e., heterogeneity about peer choices, willingness to meet guidelines, new habit acquisitions, and other variables [49–51].

Throughout the whole time, acculturation's complex operations are taking place even if the protagonists themselves do not realize it. The appearance of juvenile Hispanic gangs or *pandillas* has been explained both as a behavior copied from American-born delinquent groups deliberately devoted to the commission of criminal acts, and as a defensive reaction from young Latinos against the systematic hostility and harassments that the population's majority inflicts on them [29, 43, 52]. The adaptation levels of Hispanic immigrants vary as a result of numerous factors: from the inner disposition to adapt that each individual brings into the new reality to the degree of hospitality of rejection on the side of the host community. The former has to do with the weight and the meaning of the person's formative years, the shaping up of identity [3, 7, 12, 29, 44, 53], while the latter is a product of collective mentalities shaped up by history, culture, and socio-economic demands [23, 24].

The Clinical Scene

The clinical repercussions of acculturation operate, as well, at the individual and at the group or social level. They can impact both the physical and psychological health of the migrants, and the link with the migration process itself may not always be suspected, as the clinical manifestation may be noticed long after the arrival. The physical problems may be carried on by the immigrants (infectious or chronic diseases, allergies, fragile pre-dispositions, etc.) and triggered by environmental factors in the new setting, i.e., the impact of the "new" diet on an increased prevalence of colorectal cancer in first and second generation Hispanic immigrants [54]. Moreover, situations such as poor food intake and subsequent malnourishment and physical weakness or, on the contrary, disorganized intake, unbalanced diets, population densities in deprived urban or rural areas, and early exposition to alcohol or drugs, may contribute to obesity, hypertension, cardiovascular, respiratory or other conditions [55–58]. The lack of or difficult access to clinical services (determined by poverty, deprivation, isolation or sheer, systemic discriminatory practices by administrative border agencies [59]) may dramatically contribute to these tribulations [60, 61]. Contrarywise, a number of studies conclude that unidentified protective factors operate effectively among young adult and mature Hispanic immigrants [62, 63].

In the mental/emotional terrain, there is no doubt about the role of acculturation problems as triggering factors of a number of diagnostic conditions. It is appropriate to start with an entity included in both the WHO's International Classification of Diseases 10th. Edition (ICD-10) [64] and the APA's Diagnostic and Statistical

Manual, 5th. Edition (DSM-5) [65], under the general heading of Other Conditions that may be the Focus of Clinical Attention: Acculturation Difficulty or Acculturative Stress. The difficulties in adjusting to a new culture affect the individual's functioning and may induce clinical manifestations as varied as anxiety, fear, irritability, anger, "psychosomatic" (mostly gastrointestinal and cardiovascular), and physical (tremor, headaches, myalgias, etc.) symptoms. On occasions, and due to different severity levels of the acculturation process, or to the co-occurrence of other causal events, stress manifestations fit more precisely the picture of post-traumatic stress disorder (PTSD), with a variety of complex and severe clinical findings [66, 67].

Acculturation also plays a pathogenic role in a variety of other clinical conditions. Anxiety-related conditions, particularly generalized anxiety disorder (GAD), are probably the most frequent diagnoses, characteristic symptoms include irritability, short-temperedness, emotional lability, and insomnia. According to Hovey and Magaña [68], elevated acculturative stress, low self-esteem, ineffective social support, lack of control and decision-making capabilities, low religiosity, and high education are significantly related to high anxiety levels among Mexican immigrants. These patients may also report thinking almost obsessively about family members left behind, inability to concentrate, sudden impulses to run (at times, fantasizing with a return to their native country), mixed thoughts of violence, "anger towards God," and incessant praying and imploring for divine help. In the relationship with their physical and human surroundings, they may experience a mix of rejection, fear, avoidance, and impulsiveness. Panic-like reactions are not infrequently seen.

Closely related to anxiety-related conditions, depression in different clinical modalities is also present in this population, particularly among undocumented adolescents [69]. To typical manifestations such as sadness, low mood, profoundly nostalgic memories, social isolation, anorexia (or hyperorexia), self-critical ruminations and a sense of failure or profound discouragement, a variety of physical/somatic symptoms, i.e., headaches, shortness of breath, chest pain, or fatigability, can be added [70]. While the disruption of social networks, particularly family separation, is a strong predictor of depression [71], it is important to note, however, that in Hispanic immigrant populations, suicide or suicidal behaviors are, at times, significantly less frequent than among the majority segments: this feature is explained on the basis of religious convictions, shame in the face of suicide's meaning of defeat, pride and mutual feelings of solidarity and support [47, 72, 73].

Substance use disorders (SUDs) have also emerged with increasing force in the psychopathology of Hispanic immigrants, in spite of genuinely culture- and family-based protective resources and, ultimately, a small association between, for instance, current measures of acculturation and alcohol use among Hispanic immigrants [74]. A normative interpretation of acculturation's effects, i.e., a more intimate exposure to Anglo culture through, for instance, language use, showed stronger connections with drinking outcomes among Hispanic male and female immigrants [75].

Furthermore, the role of acculturation in these disorders has to do mainly with the opportunistic presence of narcotic traffickers in the midst of the newly arrived human groups, the offer of alcohol or substances as means of relief, the fragility and suggestibility of many newcomers, and perhaps even a sense of being "accepted" if they do what they are "invited" to do. It is clear that some segments of the migrant communities (particularly adolescents) and individual adults may show spontaneous disposition, established habits, and/or active search of drugs and alcohol before the actual transition process [76, 77].

Last but not least, acculturation may trigger true personality disorders in individuals probably predisposed by genetic-biological factors, or contribute to their appearance by inflicting powerful stressing experiences, particularly to immigrant children who, then, become vulnerable to environmental factors leading to the disorders. An important study by Breslau et al. [78] reveals the dramatic increase of behavioral disorders across generations of Mexican origin after migration to the USA: lower in the general population of the country, higher in children of Mexican-born immigrants raised in the USA, and higher still in Mexican-American children of US-born parents; nevertheless, the association with migration was markedly weaker for aggressive than for nonaggressive symptoms. Moreover, it is somewhat surprising that antisocial behaviors reach the highest levels among Hispanic adolescents and young adult immigrants who use the separation (rather than marginalization) strategy of Berry's model [20, 21, 79]. Finally, without specific epidemiological details, it seems likely that antisocial, borderline, histrionic, and passive-dependent types may be the most frequently personality disorders seen among Hispanic immigrants. Each label would have a more or less precise pathogenic chain of events leading to a final diagnosis [80, 81].

Management of Acculturation and Related Conditions

Although described primarily as a collective process, an event related to and experienced by groups of people and configured by institutional rules and norms, habits and dispositions, acculturation is also, in many ways, an individual phenomenon, a particular set of situations lived and led by a person exposed to a new, different environment, his/her new "home." Most of the studies conducted in this area, however, use a group-oriented approach, a "public health" perspective [82] together with inputs from a variety of social sciences focused, again, on collective experiences and exchanges (families, communities, neighborhoods, etc.). On its side, the individual approach seems to mostly adopt the form of "clinical cases," behavioral, mood-related or psychotic-like reports of conditions triggered and/or managed by acculturative factors [83].

The management of acculturation-related phenomena in a group context entails different foci:

1. *Educational* or formative, fundamentally oriented to basic aspects of everyday life that would make the adaptation process more manageable. Primary among them is the learning of the new language. It is well known that most Latino migrants to the USA do not speak English, so a more or less regular language-oriented training would provide them with probably the most decisive instrument for a good, constructive acculturation process [30, 42, 84]. In turn, the role

of religious organizations is extremely important both for their purpose of spiritual and emotional support and the concrete actions of solidarity and material assistance [34, 85]. On the immigrant's side, the use of integration as acculturation style serves as a protective resource [49, 86]. Needless to say, the immigrants are expected to take advantage of these opportunities, mostly implemented by school- and church-based organizations; if they do not, the result is, many times, a growing sense of isolation and self-exclusion [26, 27, 40]. Unfortunately, the numbers of those who adopt such intransigent position, dictated mainly by defensive, self-protective or anger-nourished elaborations, is not small.

2. *Social, legal, and administrative* rules, dictated by federal, state, county or local legislatures and other public organizations, supposedly in charge of an orderly location of immigrant groups, the implementation of dispositions related to personal documentation, potential job search or arrangements, living conditions, police and public order agencies, social living norms, etc. Undoubtedly well-intentioned, these tools cannot avoid a marked impersonal, bureaucratic flavor that, on occasions, may generate rejection on the side of the immigrants; their deficient use as rules to "reinforce" public order and tranquility can deviate or be interpreted, at times, as arbitrary, abusive, or discriminatory against the newcomers [5, 16, 19, 87]. The role of social media, in turn, makes communications and information-sharing a much more complex process [88].
3. *Public Health* proper, that is, mostly informative, prevention-oriented, guidelines-providing activities implemented in settings such as community centers, schools, city or municipal agencies or social organizations [89]. Entities formed by previous groups or waves of immigrants, already more or less established or integrated in the host society, occupy a front-line position in the implementation of these strategies [87, 90]. The actual procedures include public lectures, group conversations, home or neighborhood visits, city tours, telephone or online contacts. The newcomers are specifically invited to these activities, and clearly told about their purpose: make them feel comfortable enough to become quickly familiarized with their surroundings, know the rules, initiate job, school, church or other searches, and be reaffirmed in the main objectives of their move [91, 92].
4. The *political* management of the acculturation experiences falls in the hands of leaders, authorities, agencies, and parties guided by ideologies and doctrines, and implemented by public acts, manifestations, proclaims, and pronouncements. It may have, therefore, both positive and negative implications, with different measures of acceptance, popularity, persuasiveness, and effectiveness. A positive impact of the political management of acculturation is reflected in social order, equities, mutual acceptance, and pragmatic integration of the immigrant and local communities [15–19, 34, 90]. Negative impact, many times due to divisive, demagogic, confusing, changing or opportunistic positions adopted by political leaders, results in social injuries to the forthcoming members of the so-called "human capital" [93] in the form of pervasive discrimination, aversion, racism against "colored" people, and violence of all kinds (domestic, criminal, sexual and even police-led) [94–97].

The individual *clinical* management of acculturation-related conditions depends on the nature (physical/medical or mental/emotional) of the disorders. In either case, it becomes the responsibility of medical personnel in outpatient, hospital- or community-based settings: immigrants with acute or chronic physical or medical problems must be treated with the same professional dedication and promptness offered to local patients [98]. In the case of Hispanics, other than the language and communication issues [30, 42] the culturally determined attitudes towards figures of authority (i.e., health professionals) are an important factor for the success of the interactions [99, 100]. As mentioned above, probably the most frequent diagnostic conditions faced under these circumstances are nutrition-related (thinness or obesity), infectious, cardiovascular, gastrointestinal, or respiratory problems [54–58]; nevertheless, acculturation-related factors seem to accentuate ethnic and nativity-related low cardiovascular risk among foreign-born Mexican Americans when compared with non-Hispanic Whites [101]. On the other hand, lack of insurance coverage and financial restrictions among immigrants are definitely damaging factors in their overall management.

The psychological, mental, or emotional clinical conditions with acculturative stress as a point of departure require specialized care in many cases, including psychotic conditions [102]. The use of psychopharmacological agents (i.e., anxiolytics, antidepressants, major tranquilizers) is totally justified in order to treat specific target symptoms. The doses for Hispanic patients may be higher or lower than those required by Anglo-Saxon ones, depending on genetic/metabolic patterns, tolerance levels, side effect occurrences, history of previous use, etc., but the general recommendation is to start with mild to moderate amounts, and continue with periodic increments over a period of several weeks [103]. It is important to keep in mind that anxiolytic agents may induce tolerance, subsequent habituation and with further addiction and physical dependency. If there is a previous history of substance use or abuse, this chain of events becomes a complex management barrier.

These considerations make even more important the need to install a truly comprehensive treatment plan that includes appropriate psychotherapy arrangements. As acculturation is an essentially existential experience, the need to establish empathetic connections with the innermost aspects of the Hispanic patient's psyche is mandatory [96, 99, 104]. Psychotherapy, independent of school-oriented techniques, allows an open discussion of the different phases of the migration process, and particularly the impact of, until then, unknown circumstances. The approach to be used needs to emphasize the common ingredients of all psychotherapeutic procedures [105], but also the recognition of features inherent to the Hispanic condition, such as cautiousness, initial distrust, further demandingness, respect towards authority, and willingness to follow instructions once the relationship has been duly established. Some cultural psychotherapies such as the use of typical short stories or proverb interpretations [106] may be particularly effective. Biculturality in Hispanic migrants seems to predict lower HIV acquisition risk [107]. Actually, prevention and treatment services for Hispanic immigrants, aimed at increasing levels of emotional support, self-esteem, and coping skills [72, 104, 108] are critically relevant.

Discussion

Acculturation as the culminating and most powerful phase of the migration process has, in the case of Hispanic populations, a number of unique features that impose a characteristically complex structure to the bio-psycho-socio-cultural-spiritual texture of this particular aspect of their life experience [109]. In other words, it affects the total human entity of the Hispanic migrant and may become a decisive factor in the final quality of his/her adaptation to the North-American milieu. That is why, pertinent steps of the acculturative journey [110–112] must be carefully explored in any kind of interpersonal contact, administrative inquiry or clinical assessment and, if possible, integrated in an individual life report, memoir, or healthcare record.

It must also be remembered that Hispanic migrants bring with them a unique table of values, one in which family ties, religious principles and social interaction norms play crucial roles [96, 99, 110]. Acculturation in the North-American scenario truly puts such values to a critical test as adaptation, no matter how successful, entails both change and permanence: the former as a condition of co-existence with the host population's norms and lifestyles [113], some of which will be incorporated anyway; and permanence because renouncing totally to their original identity would make the Hispanic immigrant a victim of a blind, cognitive enculturation, a depersonalized, confused, self-damaging and alienated being [114]. That is why the main objective of a healthy acculturation, or of any management process to assist in it, must entail a deliberate search of harmony and balance [92, 96, 115, 116] in the context of a renewed identity [3, 34, 83, 117].

As a subject of research, acculturation offers a wide range of possibilities. Numerous studies on Hispanic immigrants in the USA clearly reflect the importance of the topic. Public Health inquiries suggest the need to refine existing tools aimed at the study of linguistic and other cultural elements looking for validity and usefulness across ethnic and subethnic groups [11, 118, 119]. The fascinating epidemiological "Hispanic Paradox" [120] constitutes a topic of almost endless possibilities in both methodological and clinical (physical and mental) areas addressing its validity and transcendency [121, 122]. Similarly, the theme of resilience and its role in the Hispanic acculturation process and experience, presents a crucial challenge both as an implicit preventive resource and an interventional strategy [123].

The risks of a failed acculturation are enormous. Whether they are originated in the personal structure of the migrant him/herself, or related to adverse features of the host community, the sequel of events in a denaturalized acculturation process go from behavioral and affective traits to well-documented clinical entities. The impact is, likewise, multifaceted because it reaches beyond the individual's life alone, to touch on his family (by his side or distant, left behind), job, finances, links to socio-political institutions, and existential outcomes. The need of a firm, constructive acculturation is not only a fact relevant to the migrant or his/her family: it is of critical relevance for the fate and stability of the world community as a whole.

Conclusions

The complexity of issues involved in the consideration and the study of acculturation has captured the attention of academicians and researchers alike, as the migratory phenomena become a decisive factor in the current realities of globalization [124, 125]. The purpose of such studies is to dissect the different components of the process from the individual and collective perspectives, in order to take away the undesirable features of a failed acculturation, and make of it, instead, a pillar of constructive coexistences. Research on acculturation is extensive and keeps growing; it has provided interesting routes of knowledge towards its true nature, its phases, ingredients and as a factor of success or failure [126]; at the same time, however, it has uncovered sources of complexity, risks, and inconveniences in clinical, demographic, socio-cultural, and financial spheres. Ultimately, research on this topic can only be thorough if and when international efforts take place in attempts to identify commonalities of constructive acculturation experiences in different parts of the world and with different populations, ethnic or cultural groups—all efforts oriented to the elimination of dissension and conflicts.

References

1. Redfield R, Linton R, Herskovits M. Memorandum on the study of acculturation. Am Anthropol. 1936;38:149–52.
2. Herskovits MJ. The significance of the study of acculturation for anthropology. Am Anthropol. 1937;39(2):259–64.
3. Schwartz SJ, Montgomery MJ, Briones E. The role of identity and acculturation among immigrant people: theoretical propositions, empirical questions, and applied recommendations. Hum Dev. 2005;304:1–30.
4. Schwartz SJ, Unger JB, Zamboaga BL, et al. Rethinking the concept of acculturation: implications for theory, measurement and health research. Am Psychol. 2010;65(4):237–51.
5. Rothe EM, Tzuang D, Pumariega AJ. Acculturation, development and adaptation. Child Adolesc Psychiatric Clin N Am. 2010;19:681–96. https://doi.org/10.1016/j.chc.2010.07.002.
6. Hughes CC. Culture in clinical psychiatry. In: Gaw AC, editor. Culture, ethnicity and mental illness. Washington D.C.: American Psychiatric Press; 1993. p. 3–42.
7. Tseng W-S. Handbook of cultural psychiatry. San Diego, CA: Academic Press; 2001.
8. Chao GT, Moon H. The cultural mosaic: a metatheory for understanding the complexity of culture. J Appl Psychology. 2005;90:1128–40.
9. Titzmann PF, Fuligni AJ. Immigrants' adaptation to different cultural settings: a contextual perspective on acculturation. Int J Psychol. 2015;50(6):407–12. https://doi.org/10.1002/ijop.12219.
10. Castro FG, Boyer GR, Balcazar HG. Healthy adjustment in Mexican American and other Latino adolescents. In: Montemayor R, Adams GR, Gullotta TP, editors. Adolescent diversity in ethnic, economic and cultural contexts. Thousand Oaks, CA: Sage; 2000. p. 141–78.
11. Thomson MD, Hoffman-Goetz L. Defining and measuring acculturation: a systematic review of public health studies with Hispanic populations in the United States. Soc Sci Med. 2009;69(7):983–91. https://doi.org/10.1016/j.socscimed.2009.05.011.
12. Suárez-Orozco C, Suárez-Orozco MM. Children of immigration. Cambridge, MA: Harvard University Press; 2001.

13. US Census. Population Reports. Available at: http://www.census.gov/population/www/index.html.
14. Escobar JI, Vega WA. Mental health and immigration's three A's: where are we and where do we go from here? J Nerv Ment Dis. 2000;188(11):736–40.
15. Fox M, Thayer ZM, Wadhwa D. Acculturation and health: the moderating role of sociocultural context. Am Anthropol. 2017;119(3):405–21. https://doi.org/10.1111/aman.12867.
16. Erten EY, van den Berg P, Weissing FJ. Acculturation orientations affect the evolution of a multicultural society. Nat Commun. 2018;9(1):58. https://doi.org/10.1038/s41467-017-02513-0.
17. Castro FG, Murray KE. Cultural adaptation and resilience: controversies, issues and emerging models. In: JW R, Zautra AJ, Hall JS, editors. Handbook of adult resilience: concepts, methods and applications. New York: Guilford; 2010.
18. Mesoudi A. Migration, acculturation and the maintenance of between-group cultural variation. PLoS One. 2018;13(10):e0205573. https://doi.org/10.1371/journl.pone.0205573.
19. Torres L, Driscoll MW, Voell M. Discrimination, acculturation, acculturative stress, and Latino psychological distress: a moderated mediational model. Cultur Divers Ethnic Minor Psychol. 2012;18(1):17–25. https://doi.org/10.1037/a0026710.
20. Berry JW. Immigration, acculturation and adaptation. Appl Psychol Int Rev. 1997;46:5–68.
21. Berry JW. Acculturation: living successfully in two cultures. Int J Intercult Relations. 2005;29:697–712.
22. Koch MW, Bjerregaard CC. Acculturation and mental health–Empirical verification of J.W. Berry's model of acculturative stress. Int J Circumpolar Health. 2004;63(Suppl 2):371–6. https://doi.org/10.3402/ijch.v63i0.17939.
23. Volpone SD, Marquardt DJ, Casper WJ, Avery DR. Minimizing cross-cultural maladaptation: how minority status facilitates change in international acculturation. J Appl Psychol. 2018;103(3):249–69. https://doi.org/10.1037/apl0000273.
24. Sangalang CC, Becerra D, Mitchell FM, Lechuga-Peña S, Lopez K, Kim I. Trauma, post-migration stress, and mental health: a comparative analysis of refugees and immigrants in the United States. J Immigt Minor health. 2019;21(5):909–19. https://doi.org/10.1007/s10903-018-0826-2.
25. Bostean G, Gillespie BJ. Acculturation, acculturative stressors, and family relationships among Latina/o immigrants. Cultur Divers Ethnic Minor Psychol. 2018;24(1):126–38. https://doi.org/10.1037/cdp0000169.
26. Lopez-Class M, Castro FG, Ramirez AG. Conceptions of acculturation: a review and statement of critical issues. Soc Sci Med. 2011;72(9):155–62. https://doi.org/10.1016/j.socscimed.2011.03.011.
27. Lara M, Gamboa C, Kahramanian MI, Morales LS, Hayes-Bautista DE. Acculturation and Latino health in the United States: a review of the literature and its sociopolitical context. Ann Rev Public Health. 2005;26:367–97.
28. Standish K, Nandi V, Ompad DC, Momper S, Galea S. Household density among undocumented Mexican immigrants in new York City. J Immigr Minor Health. 2010;12(3):10–8. https://doi.org/10.1007/s10903-008-9175-x.
29. Henriques GJB, Simon B, Ispolatov Y, Doebell M. Acculturation drives the evolution of intergroup conflict. Proc Natl Acad Sci U S A. 2019;116(28):14089–97. https://doi.org/10.1073/pnas.1810404116.
30. Shin Y, Maupome G. Segmentation of Mexican-heritage immigrants: acculturation typology and language preference in health information seeking. J Immgr Minor health. 2017;19(5):1163–73. https://doi.org/10.1007/s10903-016-0401-7.
31. Hochhausen L, Perry DF, Le H-N. Neighborhood context and acculturation among central American immigrants. J Immigr Minor Health. 2010;12(5):806–9. https://doi.org/10.1007/s10903-008-9201-z.
32. Benitez TJ, Dodgson JE, Coe K, Keller C. Utility of acculturation in physical activity research in Latina adults: an integrative review of literature. Health Educ Behav. 2016;43(3):256–70. https://doi.org/10.1177/1090198115601042.

33. Buckingham SL, Suarez-Pedraza MC. "It has cost me a lot to adapt to here": the divergence of real acculturation from ideal acculturation impacts Latinx immigrants' psychosocial well-being. Am J Orthopsychiatry. 2019;89(4):406–19. https://doi.org/10.1037/ort0000329.
34. Gibson MA. Immigrant adaptation and patterns of acculturation. Hum Development. 2001;44:19–23.
35. Dressler WW. Cultural consonance. In: Bhugra D, Bhui K, editors. Textbook of cultural psychiatry. Cambridge, U.K.: Cambridge University Press; 2007. p. 179–90.
36. Powell JA, Menendian S. The problem of othering: towards inclusiveness and belonging. Othering and Belonging, 2017. Access: otheringandbelonging.org.
37. Gonzalez A. Mexican "otherness" in the rhetoric of Mexican Americans. South Comm J. 1990;55(3):276–91. https://doi.org/10.1080/10417949009372795.
38. Mangold M, Mintz J, Javors M, Marino E. Neuroticism, acculturation and the cortisol awakening response in Mexican American adults. Horm Behav. 2012;61(1):23–30. https://doi.org/10.1016/j.yhbeh.2011.09.009.
39. Kennedy DP, Adolphs R. Social neuroscience: stress and the city. Nature. 2011;474: 452–72.
40. Tlzer EH, Yuen C, Gonzalez N, Fuligni AJ. Filling gaps in the acculturation gap-distress model: heritage cultural maintenance and adjustment in Mexican-American families. J Youth Adolesc. 2016;45(7):1412–25. https://doi.org/10.1007/s10964-015-0408-8.
41. Arévalo SP, Tucker KL, Falcón LM. Life events trajectories, allostatic load, and the moderating role of age at arrival from Puerto Rico to the US mainland. Soc Sci Med. 2014;120:301–10. https://doi.org/10.1016/j.socscimed.2014.09.040.
42. Schutt RK, Mejía C. Health care satisfaction: effects of immigration, acculturation, language. J Immigr Minor Health. 2017;19(6):1372–8. https://doi.org/10.1007/s10903-016-0409-z.
43. Smokowski PR, David-Ferdon C, Stroupe N. Acculturation and violence in minority adolescents: a review of the empirical literature. J Prim Prev. 2009;30(3–4):215–63. https://doi.org/10.1007/s10935-009-0173-0.
44. Williams LR, Rueda HA. Mexican American adolescent couples' vulnerability for observed negativity and physical violence: pregnancy and acculturation mismatch. J Adolesc. 2016;52:170–81. https://doi.org/10.1016/j.adolescence.2016.08.001.
45. Cobb CL, Xie D, Meca A, Schwartz SJ. Acculturation, discrimination, and depression among unauthorized Latinos/as in the United States. Cultur Divers Ethnic Minor Psychol. 2017;23(2):258–68. https://doi.org/10.1037/cdp0000118.
46. Germán M, Gonzalez NA, Dumka L. Familism values as a protective factor for Mexican-origin adolescents exposed to deviant peers. J Early Adolesc. 2009;29:16–42.
47. Zayas LH, Bright CL, Alvarez-Sánchez T, et al. Acculturation, familism and mother-daughter relations among suicidal and non-suicidal adolescent Latinas. J Primary Prevent. 2009;30:351–69.
48. Vermeesch AL, Stommel M. Physical activity and acculturation among U.S. Latinas of childbearing age. West J Nurs Res. 2014;36(4):495–511. https://doi.org/10.1177/0193945913507341.
49. Titzmann PF, Jugert P. Acculturation in context: the moderating effects of immigrant and native peer orientations on the acculturation experience of immigrants. J Youth Adolesc. 2015;44(11):2079–94. https://doi.org/10.1007/s10964-015-0314-0.
50. Riosmena F, Kuhn R, Jochem WC. Explaining he immigrant health advantage: self-selection and protection in health-related factors among five major national-origin immigrant group in the United States. Demography. 2017;54(1):175–200. https://doi.org/10.1007/s13524-016-0542-2.
51. Gil AS, Veg WA. Two different worlds: acculturation stress and adaptation among Cuban and Nicaraguan families. J Soc Pers Rel. 1996;13:435–56.
52. Vigil D. Barrio gangs: street life and identity in Southern California. Austin, TX: University of Texas; 1988.
53. Padilla AM, Perez W. Acculturation, social identity, and social cognition: a new perspective. Hisp J Behav Sci. 2003;25:35–55.

54. Monroe KR, Hankin JH, Pike MC, Henderson BE, et al. Correlation of dietary intake and colorectal cancer incidence among Mexican-American migrants: the multiethnic cohort study. Nutr Cancer. 2003;45(2):133–47. https://doi.org/10.1207/s15327914NC4502_01.
55. Van Wieren AJ, Roberts MB, Arellano N, Feller ER, Diaz JA. Acculturation and cardiovascular behaviors among Latinos in California by country/region of origin. J Immigr Minor Health. 2011;13(6):975–81. https://doi.org/10.1007/s10903-011-9483-4.
56. Markides KS, Lee DJ, Ray LA. Acculturation and hypertension in Mexican Americans. Ethn Dis. 1993;3(1):70–4.
57. Yoshida YX, Simonsen N, Chen L, Zhang LU, Scribner R, Tseng T-S. Sociodemographic factors, acculturation and nutrition management among Hispanic American adults with self-reported diabetes. J Health Care Poor Underserved. 2016;27(3):1592–607. https://doi.org/10.1353/hpu.2016.0112.
58. Quandt SA, Grzywacz JG, Trejo G, Arcury TA. Nutritional strategies of Latino farmworker families with preschool children: identifying leverage points for obesity prevention. Soc Sci Med. 2014;123:72–81. https://doi.org/10.1016/j.socscimed.2014.10029.
59. Sullivan MM, Rehm R. Mental health of undocumented Mexican immigrants: a review of the literature. Adv Nurs Sci. 2005;28(3):240–51. https://doi.org/10.1097/00012272-200507000-00006.
60. Adams C, Harder BM, Chatterjee A.& Mathias L.H. Healthworlds, cultural health toolkits, and choice: how acculturation affects patients' views of prescription drugs and prescription drug advertisement. Qual Health Res. 2019;29(10):1419–32. https://doi.org/10.1177/1049732319827282.
61. Allen L, Cummings J. Emergency department use among Hispanic adults: the role of acculturation. Med Care. 2016;54(5):449–56. https://doi.org/10.1097/MLR.0000000000000511.
62. John EM, Phipps AI, Davis A, Koo J. Migration history, acculturation and breast cancer risk in Hispanic women. Cancer Epidemiol Biomark Prev. 2005;14(12):2905–13. https://doi.org/10.1158/1055-9965.EPI-05-0483.
63. Riosmena F, Wong R, Palloni A. Migration selection, protection and acculturation in health: a binational perspective in older adults. Demography. 2013;50(3):1039–64. https://doi.org/10.1007/s13524-012-0178-9.
64. World Health Organization. The international classification of diseases. In: (ICD-10) classification of mental and behavioural disorders: diagnostic criteria for Research. 10th ed. Geneva: World Health Organization; 1992.
65. American Psychiatric Association. (ICD-10) classification of mental and behavioural disorders: diagnostic criteria for Research. 10th ed. Geneva: World Health Organization; 1992.
66. Rothe EM. Post-traumatic stress symptoms in Cuban children and adolescents during and after refugee camp confinement. In: Corales TA, editor. Trends in post-traumatic stress disorder research. New York: Nova Science Publishers; 2005.
67. Sondergard HP. Post-traumatic stress disorder and life events among recently resettled refugees. Stockholm: Dpt. of Public Health Sciences, Division of Psychosocial factors and Health, Karolinska Institute; 2002.
68. Hovey JD, Magaña CG. Cognitive, affective and physiological expressions of anxiety symptomatology among Mexican migrant farmworkers: predictors and generational differences. Community Ment Health J. 2002a;38(3):223–37. https://doi.org/10.1023/a:1015215723789.
69. Potochnick SR, Perreira KM. Depression and anxiety among first-generation immigrant Latino youth: key correlates and implications for future research. J Nerv Ment Dis. 2010;198(7):470477. https://doi.org/10.1097/NMD.0b013e3181e4ce24.
70. Walker JL, Ruiz RJ, Chinn JJ, Marti N, Ricks TN. Discrimination, acculturation and other predictors of depression among pregnant Hispanic women. Ethn Dis. 2012;22(4):497–503.
71. Gutierrez-Vazquez E, Flippen C, Parrado E. Migration and depression: a cross-national comparison of Mexicans in sending communities and Durham, NC. Soc Sci Med. 2018;219:1–10. https://doi.org/10.1016/j.socscimed.2018.09.064.
72. Bebbington P, Cooper C. Affective disorders. In: Bhugra D, Bhui K, editors. textbook of cultural psychiatry. Cambridge, U.K.: Cambridge University Press; 2007. p. 224–41.

73. McHale SM, Updegraff KA, Kim JY, et al. Culture orientations, daily activities, and adjustment in Mexican American youth. J Youth Adolesc. 2009;38:627–41.
74. Alvarez MJ, Frietze G, Ramos C, Field C, Zárate MA. A quantitative analysis of Latino acculturation and alcohol use: myth versus reality. Alcohol Clin Exp Res. 2017;41(7):1246–56. https://doi.org/10.1111/acer.13420.
75. Zemore SE. Re-examining whether and why acculturation relates to drinking outcomes in a rigorous national survey of Latinos. Alcohol Clin Exp Res. 2005;29(12):2144–53. https://doi.org/10.1097/01.alc.0000191775.01148.c0.
76. Prado G, Szapocznik J, Maldonado-Molina MM, et al. Drug use/abuse prevalence, etiology, prevention and treatment in Hispanic adolescents: a cultural perspective. J Drug Issues. 2008;38(1):5–36.
77. Schwartz SJ, Unger JB, Baezconde-Garbanati L, Zamboanga BL, et al. Trajectories of cultural stressors and effects on mental health and substance use among Hispanic immigrant adolescents. J Adolesc Health. 2015;56(4):433–9.
78. Breslau J, Borges G, Saito N, Tancredi DJ, et al. Migration from Mexico to the United States and conduct disorders: a cross-national study. Arch Gen Psychiatry. 2011;68(12):1284–93. https://doi.org/10.1001/archgenpsychiatry.2011.140.
79. Sobral-Fernández J, Gomez-Fraguela JA, Luengo A, Romero E, Villar P. Adolescentes latinoamericanos, aculturación y conducta antisocial. Psicothema. 2010;22(3):410–5.
80. Tsuang MT, Tohen M, Zahner GEP, editors. Textbook in psychiatric epidemiology. New York, N.Y: Wiley-Liss; 1995.
81. Alarcón RD, Foulks EF, Vakkur M. Personality disorders and culture. Clinical and conceptual interactions. New York, N.Y: John Wiley and Sons; 1998.
82. Kleinman A. Patients and their healers in the context of culture. Berkeley, CA: University of California Press; 1980.
83. Bhugra D. Acculturation, cultural identity and mental health. In: Bhugra D, Cochrane R, editors. Psychiatry in multicultural Britain. Gaskell: London, U.K; 2001.
84. Timmins CL. The impact of language barriers on the health care of Latinos in the United States: a review of the literature and guidelines for practice. J Midwif Wom Health. 2002;47(2):80–96.
85. Balidemaj A, Small M. The effects of ethnic identity and acculturation in mental health of immigrants: a literature review. Int J Soc Psychiatry. 2019;65(7–8):643–55. https://doi.org/10.1177/0020764019867994.
86. Behrens K, del Pozo MA, Großhennig A, Sieberer M, Graef-Calliess IT. How much orientation towards the host culture is healthy? Acculturation style as risk enhancement for depressive symptoms in immigrants. Int J Soc Psychiatry. 2015;61(5):498–505. https://doi.org/10.1177/0020754014560356.
87. Ramos AK, Su D, Lander L, Rivera R. Stress factors contributing to depression among Latino migrant farmworkers in Nebraska. J Immigr Minor Health. 2015;17(6):1627–34. https://doi.org/10.1007/s10903-015-0201-5.
88. Komito L. Social media and migration: virtual community. J Am Soc Inform Sci Technol. 2011;62(6):1075–86.
89. Abraído-Lanza AF, Adria N, Ambrister AM, Flórez KR, Aguirre AN. Toward a theory-driven model of acculturation in public health research. Am J Pub Health. 2006;96(8): 1342–6.
90. Kosic A, Mannetti L, Sam DL. The role of majority attitudes towards outgroup in the perception of the acculturation strategies of immigrants. Int J Intercult Rel. 2005;29(3): 273–88.
91. Zlobina A, Basabe N, Paez D, Furnham A. Sociocultural adjustment of immigrants: universal and group-specific predictors. Int J Intercult Rel. 2006;30(2):195–211.
92. Bhugra D, Desai M, Baldwin D. Focus groups: implications for primary and cross-cultural psychiatry. Prim Care Psychiatry. 1997;3:45–50.
93. Currie J. Healthy, wealthy and wise: economic status, poor health in childhood, and human capital development. NBER Working Paper No. 13987 http://www.nber.org/papers/w13987.

94. Bacallao ML, Smokowski PR. The cost of getting ahead: mexican family system changes after immigration. Fam Relations. 2007;56:52–66. https://doi.org/10.1111/j.1741-3729.2007.00439.x.
95. Carvajal SC, Hanson CE, Romero AJ, Coyle KK. Behavioral risk factors and protective factors in adolescents: a comparison of Latino and non-Latino whites. Ethnic and Health. 2002;7:181–93.
96. Alarcón RD. Mental health among Latin American migrants. In: Moussaui D, Bhugra D, Ventiglio A, editors. Mental health and mental illness worldwide. London, UK: Springer Nature; 2018. https://doi.org/10.1007/978-981-10-0750-7-7-1.
97. Bassett JF, Connelly JN. Terror management and reactions to undocumented immigrants: mortality salience increases aversion to culturally dissimilar others. J Soc Psychol. 2011;151(2):117–20. https://doi.org/10.1080/00224540903365562.
98. Lolas F. Ethics in psychiatry: a framework. World Psychiatry. 2006;5:185–7.
99. Alarcón RD. Identidad de la Psiquiatría Latinoamericana. Voces y exploraciones en torno a una ciencia solidaria. México D.F.: Siglo XXI Editores; 1990.
100. Salgado H, Castañeda SF, Talavera GA, Lindsay SP. The role of social support and acculturative stress in health-related Quality of Life among day laborers in Northern San Diego. J Immigr Minor Health. 2012;14(3):379–85. https://doi.org/10.1007/s10903-011-9568-0.
101. Kershaw KN, Greenlund K, Stamler J, Shay CM, Daviglus ML. Understanding ethnic and nativity-related differences in low cardiovascular risk status among Mexican-Americans and non-Hispanic Whites. Prev Med. 2012;55(6):597–602. https://doi.org/10.1016/j.ypmed.2012.09.019.
102. Weisman de Mamani A, Weintraub MJ, Maura J, et al. Acculturation styles and their associations with psychiatric symptoms and quality of life in ethnic minorities with Schizophrenia. Psychiatry res. 2017;255:418–23. https://doi.org/10.1016/j.psychres.2017.06.074.
103. Ayuso-Gutierrez JL. Tratamiento farmacológico de Trastornos Ansiosos. In: Alarcón RD, Chaskel R, Berlanga C, editors. Psiquiatría, 4th. Ed. Lima, Perú: Fondo Editorial Universidad Peruana Cayetano Heredia; 2019. p. 594–600.
104. Cobb CL, Branscombe NR, Meca A, Schwartz SJ, et al. Toward a positive psychology of immigrants. Perspect Psychol Sci. 2019;14(4):619–32. https://doi.org/10.1177/1745691619825848.
105. Frank JD, Frank JB, editors. Persuasion and healing: a comparative study of psychotherapy. 3rd ed. Baltimore, MD: Johns Hopkins University Press; 1991.
106. Alarcón RD, Frank JB, Williams M. Cultural dynamics in psychotherapy and cultural psychotherapies: ingredients, processes and outcomes. In: Alarcón RD, Frank JB, editors. The psychotherapy of hope. The legacy of persuasion and healing. Baltimore, MD: Johns Hopkins University Press; 2012. p. 281–309.
107. Kang S-Y, Deren S, Mino M, Cortés DE. Biculturality and HIV-risk behaviors among Puerto Rican drug users in new York City. Subst Use Misuse. 2009;44(4):578–92. https://doi.org/10.1080/10826080802487341.
108. Hovey JD, Magaña CG. Exploring the mental health of Mexican migrant farmworkers in the Midwest: psychosocial predictors of psychological distress and suggestions for prevention and treatment. J Psychol. 2002b;136(5):493–513. https://doi.org/10.1080/00223980209605546.
109. Killian C. Integrative perspectives on acculturation. Am Psychol. 2011;66(2):154–7. https://doi.org/10.1037/a0022242.
110. Glazer S, Roach KN, Carmona C, Simonovich H. Acculturation and adjustment as a function of perceived and objective value congruence. Int J Psychol. 2018;53(Suppl. 2):13–22. https://doi.org/10.1002/ijop.12554.
111. Fox M, Thayer ZM, Ramos IF, Meskal SJ, Wadhwa PD. Prenatal and postnatal mother-to-child transmission of acculturation's health effects in Hispanic Americans. J Women's Health. 2018;27(8):1054–63. https://doi.org/10.1089/jwh.2017.6526.
112. Alderete E, Vega WA, Kolody B, Aguilar-Gaxiola S. Effects of time in the United States and Indian ethnicity on DSM-III-R psychiatric disorders among Mexican Americans in California. J Nerv Ment Dis. 2000;188(2):90–100. https://doi.org/10.1097/00005053-200002000-00005.

113. Andreeva DA, Unger JB. Determinants of host society acculturation and its relationship with health behaviors and outcomes: a new research and intervention framework. J Immigr Minor Health. 2015;17(5):1420–6. https://doi.org/10.1007/s10903-014-0104-x.
114. Sun S, Hoyt WT, Brockberg D, Lam J, Tiwari D. Acculturation and enculturation as predictors of psychological help-seeking attitudes (HSAs) among racial and ethnic minorities: a meta-analytic investigation. J Couns Psychol. 2016;63(6):617–32. https://doi.org/10.1037/cou0000172.
115. Chakraborty BM, Chakraborty R. Concept, measurement and use of acculturation in health and disease risk studies. Coll Antropol. 2010;34(4):1179–91.
116. Riedel J, Wiesmann U, Hannich H-J. An integrative theoretical framework of acculturation and Salutogenesis. Int Rev Psychiatry. 2011;23(6):555–64. https://doi.org/10.3109/09540261.2011.637912.
117. Bhugra D. Migration, distress and cultural identity. Brit Med Bull. 2004;69:129–41.
118. Rogler LH, Cortes DE, Malgady RG. Acculturation and mental health status among Hispanics. Convergence and new directions for research. Am Psychol. 1991;46(6):585–97. https://doi.org/10.1037/0003-066x.46.6.585.
119. McCabe BE, Vermeesch AL, Hall RF, Peragallo NP, Mitrani VB. Acculturation and the Center for Epidemiological Studies-Depression Scale for Hispanic women. Nurs Res. 2011;60(4):270–5. https://doi.org/10.1097/NNR.0b013e318221b8dc.
120. Markides KS, Eschbach K. Aging, migration and mortality: current status of research on the Hispanic paradox. J Gerontol. 2005;60(2):S68–75.
121. Bostean G. Does selective migration explain the Hispanic paradox? A comparative analysis of Mexicans in the U.S. and Mexico. J Immigr Minor Health. 2013;15(3):624–35. https://doi.org/10.1007/s-10903-012-9646-y.
122. Teruya SA, Bazargan-Hejazi S. The immigrant and Hispanic paradoxes: a systematic review of their predictions and effects. Hisp J Behav Sciences. 2013;35(4):486–509.
123. Serafica R, Lekhak N, Bhatta T. Acculturation, acculturative stress and resilience among older immigrants in the United States. Int Nurs Rev. 2019;66(3):442–8. https://doi.org/10.1111/inr.12522.
124. Stiglitz J. Globalization and its discontents. London, U.K.: Penguin Books; 2002.
125. Chorost M. World wide mind: the coming integration of humanity, machines and the internet. New York, NY: Free Press; 2011.
126. Koneru VK, Weisman de Mamani AG, Flynn PM, Betancourt H. Acculturation and mental health: current findings and recommendations for future research. Appl Prev Psychol. 2007;12:76–96.

8 Suicidal Thoughts and Behaviors in Hispanic and Latino Communities

Danielle Torres, Sharely Fred-Torres, Elizabeth Soto, and M. Mercedes Perez-Rodriguez

Abbreviations

SI:	Suicidal ideation
SA:	Suicide attempt
ED:	Emergency department
US:	United States
NSSI:	Non-suicidal self-injury

Introduction: Suicidal Thoughts and Behaviors

Suicide is one of the leading causes of death worldwide, accounting for almost one million deaths per year [1]. It is a highly complex global health issue with marked differences across gender, age, race, ethnicity, socioeconomic status, and culture. Despite the severity of this problem, progress in suicide prevention is impeded by the lack of consistent data on how social, economic, political, and cultural factors influence suicidal behavior among diverse populations [2]. A stronger understanding of specific risk factors would facilitate more efficient and effective identification of high-risk individuals, and catalyze appropriate intervention and treatment [3].

Danielle Torres and Sharely Fred-Torres contributed equally with all other contributors.

D. Torres · S. Fred-Torres · M. M. Perez-Rodriguez (✉)
Psychiatry Department, Icahn School of Medicine at Mount Sinai (ISMMS), New York, NY, USA
e-mail: mercedes.perez@mssm.edu

E. Soto
The City University of New York (CUNY), New York, NY, USA

R. Castilla-Puentes, T. Falcone (eds.), *Mental Health for Hispanic Communities*, https://doi.org/10.1007/978-3-031-13195-0_8

Suicide-related behaviors are classified by several distinct but overlapping categories ranging from completed suicide to non-suicidal self-injury (NSSI) [2, 4]. Traditional process models conceptualize the suicidal process as a sequence of thoughts and behaviors that occur in order of increasing severity beginning with passive suicidal ideation, followed by active suicidal ideation, development of a suicidal plan, and finally, suicide attempt, which may lead to death [5].

Passive Suicidal Ideation Among Hispanic and Latino Communities

Passive suicidal ideation (SI) is defined as the feeling that one's life is not worth living or fighting for, or that one would be better if dead. Passive SI is associated with increased lifetime risk of suicide attempt [6], and in fact it has been suggested that passive and active SI are equally related to increased suicide risk [5]. Passive SI can arise in the setting of stressful life events, such as familial, financial, or legal issues [7].

Research suggests that the suicidal process is not linear but rather characterized by many possible pathways. One possible pathway is direct progression from passive suicidal ideation to suicide attempt. Therefore, although active suicidal ideation receives significant attention in research, effective suicide prevention requires recognition of passive suicidal ideation as an important element of the suicidal process [5]. This is particularly important given that high rates (20–30%) of passive suicidal ideation have been reported in Hispanic populations [6].

Active Suicidal Ideation Among Hispanic and Latino Communities

Prevalence

Active SI refers to thoughts about taking action to end one's own life [3]. The lifetime prevalence of suicidal ideation among Hispanics in the United States has been reported to be 10.2% [8]. Data from the National Latino and Asian American Study (NLAAS) found no statistically significant differences in prevalence of suicidal ideation across the Latino subgroups of Puerto Rican, Cuban, Mexican, and Other Latino [8]. Interestingly, although female Hispanics were found to have higher rates of suicide attempt, they were not at greater risk for suicidal thoughts than male Hispanics [8].

Throughout the last decade, suicidal ideation among US Hispanics has increased significantly across generations. These results have been attributed to acculturative stress, degradation of individual values, and the weight of social bonds. The prevalence of SI among Latinos varies across gender differences, sexual orientation, and socio-economical differences across heterogeneous Hispanic cultures [9]. Although Hispanics are at greater risk for suicidal behavior compared to other ethnic groups, they have been reported to have a lower risk of death by suicide compared to non-Hispanics, which is known as the "Latino paradox" [10].

Risk Factors

Comorbid psychiatric illness has been found to be highly correlated with suicidal ideation (those who answered yes to having ever seriously thought about committing suicide) among Latino populations, even after adjusting for age, gender, and language [8]. Specifically, Latinos who had depression, anxiety, or substance use disorders were more likely to endorse suicidal ideation than those who did not [8]. Data from the National Latino and Asian American Study (NLAAS) found that comorbid psychiatric disorder, English proficiency, family conflict, having a US-born parent, and not attending church were all positively correlated with suicidal ideation among Latinos [8].

Identification of risk factors that are relevant to Hispanic and Latino communities has contributed to improved prevention of suicide in these communities [11]. The variation of suicide risk factors among the Latino community highlights the great need for culturally relevant suicide research for this specific population.

Suicide Plans Among Hispanic and Latino Communities

Cross-national studies suggest that having a suicide plan significantly increases the risk of SA among those with SI (those who said yes to seriously thinking about committing suicide) [7]. This finding persists in Latino populations, wherein suicidal ideation, plans, and attempts are all correlated with one another [12].

Prevalence

The lifetime prevalence of suicidal plans among adults in the United States is 3.9% [13]. Many studies [14, 15] though not all [16–18] suggest that rates of suicide plans are higher in Hispanics than non-Hispanic whites and African Americans. Epidemiologic data corroborates this finding, as Hispanic youth have been found to be more likely to develop suicidal plans than their non-Hispanic white and African American peers [19–21].

Risk Factors, Gender, and Comorbidities

Risk factors for suicidal behaviors, such as developing suicide plans, include comorbid psychiatric illness and being unmarried, young, or female [13]. While males are more likely to complete suicide, females are at greater risk for nonfatal suicidal behaviors including suicidal ideation, attempt, and plan [13, 15, 22–24]. Young female Latinas appear to be particularly high risk for suicidal behavior when compared to non-Hispanic white females and Hispanic males [12, 25]. One study of 650 Latina adolescents found that rates of SI (24%), suicide plans (17%), and suicide attempts (12%) were all higher than the national averages [12].

Table 8.1 Key points: suicide plans

• Hispanic youth more likely to develop suicide plans than non-Hispanic white and African American youth. • Young female Latinas at higher risk for non-fatal suicidal behavior compared to other ethnic and gender subgroups. • Depression is a prevalent risk factor for suicide plans among Latinos.

Table 8.2 Risk factors for suicide plans among Latinos

• Comorbid psychiatric illness including depression • Female gender • Younger age • Being unmarried • Engaging in bullying against others (for females)

Depression may be a particularly salient risk factor in Hispanic communities, as studies suggest that rates of depression are higher among Hispanic adolescents than non-Hispanic white and black adolescents [21, 26, 27]. More specifically, young Hispanic females have higher rates of depression than their Hispanic male counterparts [16, 28–31]. Studies involving Latina adolescents suggest that depressive symptoms are associated with suicidal ideation, plans, and attempts, in addition to a higher propensity to engage in bullying [12]. More specifically, young Latinas who bullied others were more likely to have developed suicide plans [12] (Tables 8.1 and 8.2).

Suicide Attempts among Hispanic and Latino Communities

Prevalence

The relationship between ethnicity and risk for SA is not well established. Some studies suggest that Hispanics have lower rates of suicide attempts than non-Hispanic whites [24, 32–34], while others posit that there is no significant difference [8, 15, 35]. Recent reports suggest that the rate of past-year suicide rates among Hispanic adults is slightly greater than that of US adults overall [36]. Epidemiologic data from 2018 found that the rate of past-year suicide attempt in those older than 18 was 0.8% among Hispanics and Latinos compared to 0.6% among US adults overall [36].

The inconsistency of these findings may be attributable to the highly heterogeneous nature of Hispanic populations [8, 9]. To this point, the prevalence of SA appears to vary within Latino subgroups. Multiple studies report higher lifetime risk of suicide attempt in Puerto Ricans than in other ethnic subgroups [2, 8, 20, 37–39]. Reviews of epidemiologic data provide greater insight. After stratifying by age and sex, the National Epidemiological Survey on Alcohol and Related Conditions (NESARC) from 2001-2002 found that Puerto Rican women ages 45–64 were the only high-risk group identified, though their rates of lifetime suicide attempt were not statistically different from that of white women of the same age [40].

Recent reports from the CDC suggest that Hispanic youth are at greater risk for SA when compared to youth nationwide. Epidemiologic data from 2017 reports that the rate of past-year SA among Hispanic high schoolers is 8.2%, a sizable increase from 7.4% in high schoolers overall [41]. Numerous studies corroborate this report, noting elevated rates of suicide attempt in Hispanic youth compared to non-Hispanic white peers [14, 19, 42, 43].

Risk Factors and Comorbidities

Risk factors for SA, including previous attempt, comorbid psychiatric illness, low socioeconomic status, substance use, stressful life events, being unmarried, and being female [13, 44, 45] also apply to Hispanic populations [8, 16, 17, 34, 38, 46, 47]. However, several risk factors are more prevalent among Latino populations. For instance, national reports from 2018 suggest that rates of depression are higher among Hispanic adults (8.2%) compared to non-Hispanic white (7.9%) and non-Hispanic Asian adults (3.1%) [48]. Higher rates of depression in specific Hispanic subgroups, including Hispanic youth [12, 21, 26, 27] and Puerto Ricans [9, 33, 38] may in part explain their higher rates of SA. Prevalence of low socioeconomic status is also greater among Hispanics compared to non-Hispanic whites [49]. Hispanics on average have lower incomes, lower levels of educational attainment, and higher rates of poverty compared to non-Hispanic whites [49].

Studies have found that among Hispanic populations, greater acculturation, which is the process by which immigrants adopt a dominant society's cultural practices, is associated with elevated risk for SA [34, 50]. Among Latino populations, various aspects of acculturation, including language proficiency and birthplace, have been found to confer increased risk for the development of psychiatric disorders [47, 51, 52] and suicide attempts [8, 53]. A study of Mexican-American teens found that greater acculturation was associated with elevated suicidality but only in combination with depression and poor self-esteem [30]. Diminished cultural and religious protective factors as a result of acculturation may contribute to elevated suicide risk [8, 54].

Gender Differences

Studies suggest that the rates of SA may be two to three times greater in females versus males [45], a finding that appears to be consistent in Hispanic populations [8, 43, 45]. Young Hispanic females are at particularly high risk [12, 43], as nearly one in five Hispanic girls in the US attempts suicide [43, 55, 56]. The high rates of depression among Hispanic women may be driving the data on suicidality. Research suggests that Hispanic women experience depression at twice the rate of Hispanic males, and they are also more likely to be depressed than Caucasian and African American women [57]. Hispanic norms around culture, gender, and family have also been discussed as contributing factors [12, 43] (Tables 8.3 and 8.4).

Table 8.3 Key points: suicide attempt

- Rate of past-year suicide attempt higher in Hispanic adults (0.8%) compared to US adults overall (0.6%).
- Rate of past-year suicide attempt higher in Hispanic high schoolers (8.2%) than US high schoolers overall (7.4%).
- Rates of suicide attempt higher among Hispanic females than Hispanic males.

Table 8.4 Risk factors for suicide attempt among Latinos

- Prior suicide attempt
- Comorbid psychiatric illness including depression
- Female gender
- Low socioeconomic status
- Being unmarried
- Stressful life events
- Greater degree of acculturation

Non-suicidal Self-Injury (NSSI) Among Hispanic and Latino Communities

Non-suicidal self-injury (NSSI) is defined as the act of injuring oneself without conscious suicidal intent, i.e., without intent to die [58]. NSSI peaks in adolescence to early-adulthood [59, 60] and typically plays a role in emotional regulation [61–64].

Prevalence

The prevalence of NSSI among adolescents may be as high as 13–28% within the community [65] and 68% among psychiatric inpatients [66]. The relationship between race or ethnicity and risk of NSSI is not well established. While some studies purport that non-Caucasian adolescents have lower rates of NSSI compared to white teens [46, 67], other studies have found no significant differences in prevalence of NSSI between these two ethnic groups [68–70].

The reported risk of NSSI in Latino populations is variable. Some studies suggest that rates of NSSI are lower in Hispanic youth compared to Caucasian counterparts [71–74]. Other studies suggest that prevalence of NSSI among Hispanic students is not significantly different from that of Caucasian students [20, 63, 64, 75, 76]. Still, other studies have reported rates of NSSI among Hispanics to be higher than those reported among white individuals [73]. These inconclusive findings may be attributable in part to inadequate representation of Hispanic populations in studies assessing NSSI [71].

Risk Factors, Comorbidities, and Gender Differences

Risk factors for NSSI include childhood trauma and sexual abuse [77–82], familial conflict [83], and reduced ability to cope with stress [84–86]. Researchers have proposed various theoretical explanations for the link between childhood abuse and neglect and NSSI, including disruptions of secure attachment between children and their caregivers [75, 87]. Other risk factors for NSSI, which have found to be relevant in populations of ethnically diverse adolescents, include dissociative symptoms and borderline personality traits [46, 71, 75].

Among a cohort of Latina adolescents, those who endorse NSSI were more likely to engage in bullying [88]. Also among Latina adolescents, those who perceive negative feelings as alterable are more likely to engage in NSSI, utilizing it as a tool for emotional regulation, whereas those without feelings of control are more likely to attempt suicide [88]. Nevertheless, Latina adolescents represent a particularly vulnerable group, as even those who perceive negative feelings as alterable are still at risk for suicidal behaviors.

Despite the fact that males are more likely to die by suicide, females are more likely to engage in NSSI, in addition to other non-fatal suicidal behaviors [15, 22–24, 64, 86, 89–92]. The overall female to male ratio of those who engage in NSSI is approximately 3:1 to 4:1 [93]. While some studies have highlighted the high rates of NSSI among Latina adolescents [43], others suggest that the rates of NSSI may be comparable among both female and male Hispanics [94] arguing that the gender difference may actually be attenuated in non-Western populations [95].

Risk factors for NSSI among Hispanic individuals may vary based on gender. One study involving 1651 Hispanic adolescents found that discrimination-related stress was associated with self-harm behavior among Hispanic males [94]. Among Hispanic females, immigration, acculturation, and family drug related stressors were associated with self-harm behavior [94] (Tables 8.5 and 8.6).

Table 8.5 Key points: Non-suicidal self-injury among Hispanic and Latino communities

- Prevalence of NSSI in Latino populations is not well established.
- NSSI is more prevalent among females than males.
- Risk factors for NSSI among Latino communities may differ by gender.
- Ethnic identity and belonging may be protective against NSSI in Latino populations.

Table 8.6 Risk factors for NSSI among Latinos

- Female gender
- Childhood trauma, including sexual abuse
- Family conflict
- Poor ability to cope with stress
- Dissociative symptoms
- Borderline personality traits
- Engaging in bullying behavior
- Stress related to discrimination (particularly in males)
- Stress related to immigration, acculturation, and drugs (particularly in females)
- Stress related to family relations (particularly in females)

Effect of Acculturation, Discrimination, and the Stress of Migration on Suicidal Thought and Behaviors Among Hispanic and Latino Communities

Acculturation has been associated with an increased risk of suicidal ideation [96], attempts [97], and lifetime prevalence of psychiatric disorders [98] among Hispanics. Through acculturation immigrants may experience cultural estrangement and discrimination that contribute significant psychological stress, known as the acculturative stress hypothesis. Immigration, by definition, imposes a new set of cultural expectations, and thus, a new identity, on minority groups. The process of redefining one's values, such as those tied to the Hispanic ethnic identity, could contribute to a greater sense of alienation [99] that increases the risk of suicide in this population.

A multi-site study demonstrated that the rate of 30-day SI was significantly associated with a lower sense of belonging, as captured by the Cultural Identity Scale for Latinos [6]. Questions from this scale inquire about identification with, closeness to, and amount of time spent with people of the same ethnic background, all of which are disrupted with greater integration into a foreign culture. Another study of over 4000 immigrants of diverse ethnic backgrounds demonstrated that longer time spent in the United States is associated with greater suicidality [100]. Specifically, this cross-sectional study demonstrated that Latinos who have lived in the United States for greater than 20 years have almost 3 times greater risk of SI compared to those who have been in the United States for less than 5 years, suggesting that increased duration of acculturation may result in a greater accumulation of stressors linked to suicidality. This finding was replicated by another study that found immigrants to have a lower lifetime prevalence of any psychiatric disorder compared to Mexican Americans born in the United States and the US population as a whole [101].

One major stressor that results from acculturation is separation and disruption of existing social networks, such as family, that could protect against suicidality. A study showed that with greater acculturation, perceived obligation to family, tied to the previously mentioned Hispanic value known as familism, decreased among Hispanics [102]. A separate study of almost 1000 ethnically diverse adults demonstrated that familial acculturative stress predicted almost 2 times greater odds of a previous SA, and that in Latinos, social acculturative stress specifically was associated with almost 3 times higher odds of a SA [103].

A large, nationally representative study of over 6000 Hispanic adults in the United States from the National Epidemiological Survey of Alcohol and Related Conditions (NESARC) reported that several factors linked to greater acculturation, such as younger age at migration, longer time in the United States, higher degree of English language orientation, lower composition of Hispanic social network, and lower identification with the Hispanic ethnic identity were all associated with increased lifetime risk for SI and SA . This study also specifically highlighted that perceived discrimination was linked to greater risk of lifetime SI and SA [53] which has been confirmed by several recent large population studies [104]. A multi-site study from cities in the United States and Europe highlighted that perceived exposure to discrimination due to inability to speak English or as a result of being Latino was significantly associated with 30-day SI in Hispanics [6]. In one study, discrimination was associated with as high as 3 times risk of SA among Latinos [103].

Suicidal Behavior in US-born Versus Foreign-born Hispanics

One's propensity to experience stress from acculturation may be dependent on birthplace, as foreign-born individuals who've spent less time interacting with individuals of their immigrant country appear to be protected from psychological morbidity. One possible explanation is that country of origin influences the extent to which native culture is preserved among Hispanic groups. Alternatively, regardless of national Hispanic origin, rates of suicidality may be disproportionately elevated for US-born Hispanics who've experienced greater acculturation.

A nationally representative study of over 8000 Mexican individuals in the United States and Mexico demonstrated that Hispanics who have immigrated to or have a family member in the United States have higher rates of suicidal behavior compared to Hispanics who have not immigrated to the United States or do not have a family member in the United States [105]. These researchers and others also examined rates of SI and SA in Hispanics born in the United States, compared to those who were not. Several studies have reported significantly higher rates of SI and SA in US-born Hispanics compared to non-US-born Hispanics [105, 106].

One study specifically reported almost 3x greater suicidal thoughts in Mexican Americans born in the United States (13%), compared to those born in Mexico (4.5%), even after adjusting for age and gender-specific rates of suicidality. Compared to non-Hispanic whites born in the United States, this study demonstrated that the rate of suicidal thoughts remained significantly lowest among Mexican Americans born in the Mexico (1.6%) and highest in Mexican Americans (4.8%) and non-Hispanic whites (4.4.%) born in the United States, suggesting that regardless of national Hispanic origin, rates of suicidality are disproportionately elevated for Hispanics born in the United States [34]. In fact, data from a large population-based study of immigrants from several geographic regions across Latin America, Europe, and Asia reported lower rates of suicidality in immigrants in the United States compared to US-born individuals [107], further suggesting that the higher rates of suicidality in US-born minority groups may not be unique to Latin American immigrants.

In comparison to Hispanics of Mexican and Cuban descent, Hispanics of Puerto Rican descent have the highest frequency of depression [33] and suicide attempts [38, 39], even when accounting for demographic, psychiatric, and sociocultural factors [8]. One study also reported that compared to non-Hispanic whites, Puerto Ricans have higher rates of suicide attempts, particularly Puerto Rican middle-aged women [40]. Given that Puerto Rican culture consists of a hybrid of American and Latin American traditions, this data provides further evidence for acculturation as a risk factor, and conversely, for the presence of a well-defined culture as a buffer against suicidal behavior. This data also suggests that suicidal behavior in Hispanic populations varies across nationalities.

Table 8.7 Key points: immigration, acculturation, and discrimination

- Difficulties with acculturation, such as cultural estrangement and discrimination, may lead to significant psychological stress.
- Higher rates of acculturation from greater time spent in a foreign country may be a unique risk factor for suicidality among US-born Hispanics, compared to foreign-born.

These higher trends of suicidal behavior in US-born Latin Americans compared to foreign-born Latin Americans are evident prior to immigration [105] and can be seen as early as young adulthood. One study demonstrated that second-generation youth (US-born Hispanics with immigrant parents) may be almost 3 times more likely to attempt suicide than first-generation (foreign-born) youth. Later generations of US-born Hispanic youth with US-born parents have even higher rates of suicidality, almost 4 times that of first-generation youth, thought to be partially mediated by substance use [108]. A larger study similarly reported that the lower prevalence of depression among Mexicans who had arrived in the United States at a later age, compared to those who immigrated as children, was no longer significant when accounting for contextual factors, such as integration to a disadvantaged neighborhood [109].

There are studies that have reported opposite trends of higher suicidality in immigrants compared to US-born Latin Americans. A frequently cited study reported higher risk of suicide in immigrants compared to native-born counterparts, specifically in areas with smaller immigrant populations. On the other hand, US-born Hispanics had a greater risk of suicide in areas with large immigrant populations [99]. These findings suggest that feeling as if one belongs to a minority or "other" group may pose a greater burden to mental health than country of origin itself.

Given the significant diversity of culture and experiences with acculturation among Hispanic groups, researchers who used data from one of the largest population-based survey studies of Latino and Asian Americans in the United States known as the National Latino and Asian American Study (NLAAS) argue that a contextual approach has to be taken by acculturation studies to appropriately understand the impact of cultural integration in Hispanic groups [109, 110] (Table 8.7).

Protective Factors Among Hispanic and Latino Communities

Despite the fact that Hispanics are at greater risk for suicidal behavior, they have been found to be at lower risk to die by suicide. Several reasons have been postulated for this so-called "Latino paradox" [10]. As demonstrated by a study that administered the Reasons For Living Inventory (RFLI) to Latinos and non-Latinos with various psychiatric disorders, Latinos scored higher on subscales assessing survival and coping beliefs, responsibility to family, and moral objections to suicide, which are associated with decreased SI and less lethal attempts [54]. Taking all of this together, it could be that one protective factor against suicidality is perception of belonging to a particular cultural group.

Several studies have found that ethnic belonging is protective against suicidal behavior [109, 111] including NSSI [76]. In a cohort of Hispanic male university students, strong Mexican ethnic identity was found to be protective against NSSI [20]. As suggested by Wadsworth and Kubrin [99], an explanation for this could be that a strong sense of cultural identity and integration, tied to Hispanic values such as familism and religiosity, may attenuate feelings of alienation, reducing the risk of suicide.

Familism

The tendency to place the needs of the family before the needs of the individual is a common Hispanic social value known as familism. According to Sabogal et al., familism specifically refers to "a strong identification and attachment of individuals and their families ... and strong feelings of loyalty, reciprocity and solidarity among members of the same family," which has been noted to be a core characteristic of Hispanic culture, regardless of ethnic origin [102]. Several studies have demonstrated that responsibility to family [54] and particularly for youth, close, positive relationships with family members [112–114] protect against suicidality among Hispanics.

Religious/Moral Beliefs Against Suicide

Similarly, given high rates of religiosity in Hispanic populations [115], several studies have focused on the protective role of religion in suicidal behavior. Researchers have demonstrated that greater influence of religion as demonstrated by greater church attendance, for example, serves a protective factor against SI in Latin American immigrants [32, 34, 116]. Religion appears to uniquely influence US-born Hispanics compared to foreign-born Hispanics, as demonstrated by a study that showed that US-born Hispanics benefit from participation in religious communities, regardless of denomination, while foreign-born Hispanics benefit specifically from Catholicism [106]. One reason could be that compared to other religious groups, Catholics tend to show greater levels of social integration and stronger ties with the community, which may be particularly important for foreign-born individuals. This stronger link to the community has been found to result in lower suicide rates compared to other religious groups [117, 118].

A well-known hypothesis for why religion could protect against suicidality is the community support theory, which suggests that belonging to homogenous religious communities provides individuals with "networks" of emotional and social support which protect against suicide [117]. Another theory is the community norms theory, which suggests that rather than provide individuals with social support, religion protects against suicidality by virtue of condemning suicidality as bad and unacceptable behavior [119]. The majority of research supports the community support theory, as early on, Durkheim suggested that the level of integration to a religious

Table 8.8 Key points: protective factors among Hispanic and Latino communities

- A strong sense of cultural identity may attenuate feelings of alienation and protect Hispanic immigrants against suicidality.
- Strong ties to Hispanic values, such as familism and religiosity, may protect against suicidality, partially through the provision of social support.

community is inversely proportional to suicide rates [120]. Similarly, Wadsworth and Kubrin point out "when immigrants arrive in a new city without emotional support systems…or ability to effectively connect with friend of kinship networks, they are more likely to experience alienation and loneliness, which in extreme cases may result in suicide" [99] (Table 8.8).

Evidence-Based Interventions to Prevent Suicidal Thoughts and Behaviors Among Hispanic and Latino Communities

Establishing effective interventions against suicidality in Hispanics is particularly important given unique barriers to mental health treatment noted in this population. Several studies have demonstrated that compared to non-Hispanics, Hispanics are less likely to utilize mental health services [121–124]. Cultural and immigration factors appear to influence the likelihood that Hispanics will seek mental health services, as one study demonstrated that Puerto Ricans and US-born Latinos are more likely than foreign-born Latinos to seek mental health treatment [109], and another highlighted lower acculturation as a risk factor for not seeking mental health services [125]. There are financial and nonfinancial reasons for this phenomenon as well. A major barrier is that compared to white counterparts, Hispanics in the United States have a lower level of educational achievement compared to non-Hispanic whites, which results in lower income and ability to pay for healthcare [126, 127]. A large population-based study highlighted that in the United States, compared to non-Hispanics, a greater proportion of Hispanics are employed in low-wage sectors in which employers fail to provide health insurance. Uninsured Hispanics are less likely to have a regular source of healthcare compared to Hispanics with private health insurance [128]. Furthermore, undocumented immigrants who face additional concerns, such as mistrust of providers due to fear of deportation or discrimination [129], have been noted to utilize health services significantly less than their US-born counterparts [130]. Interventions aimed at reducing this gap include greater educational guidance from an early age, identifying Hispanics who may need additional guidance in school or financial assistance to complete educational training required for employment in higher-wage sectors. Furthermore, community outreach programs that may link Hispanic migrants to appropriate follow-up care may facilitate access to healthcare in this population [131].

Other reasons that Hispanics are less likely to access mental health care include poor health literacy, stigma, and negative attitudes towards mental health care, even from healthcare providers [132]. One study assessed the impact of an educational intervention for Emergency Department staff on adherence to outpatient mental health treatment. This intervention trained ED providers to properly identify suicidal behavior, reinforce the importance of outpatient treatment, and provide an initial family-based therapy session in the ED when a patient presented with suicidal behavior. Results highlight the effectiveness of this intervention in decreasing depression and increasing outpatient follow-up in a group of 140 Latina adolescents, in line with the US Department of Health and Human Services goal to ensure adequate outpatient treatment as a means of preventing suicidality [133]. Another RCT study that examined the effect of educational brochures on health literacy, stigma, and attitudes towards seeking mental health treatment found that while this form of passive psychoeducation was effective in increasing knowledge on suicidal behavior, it was not effective in reducing stigma or increasing help-seeking attitudes in a group of first-generation Hispanic immigrants mostly of Mexican origin [134]. Thus, interventions specifically targeting stigma towards mental health treatment have great potential in increasing access to care among Hispanics.

Furthermore, culturally sensitive mental health resources need to be made more available to decrease discrimination and stigma among this population [53]. At present, language in itself is a barrier to Hispanics seeking mental health treatment [135]. Future efforts to address this barrier should focus on incorporating validated Spanish medical resources into clinics and incentivizing physicians to become certified in medical Spanish or use interpreters in their provision of healthcare [136].

Given ample evidence that acculturation poses an increased risk for suicidality via disintegration of cultural values, other researchers have focused on culturally based preventive interventions as a means of reducing suicide rates in Hispanics. One parent-centered intervention aimed at decreasing adolescent drug use and unsafe sexual activity, known as Familias Unidas, shows potential in fostering familism as demonstrated by greater parent–adolescent communication [137] and efficacy in reducing internalizing symptoms [137, 138] that are linked to suicidality [139] in this population. This intervention was noted to be particularly effective for families with high stress and low levels of social support [139]. Findings of greater church attendance linked to lower rates of suicidality [34] and of interventions aimed at fostering positive relationships with school adults as buffers against suicidality [140] imply that the Catholic church and high schools could be a focus of community-based interventions [34].

To this date, no studies have assessed the effectiveness of interventions intended to augment cultural engagement in this group. Nonetheless, a group of researchers recently proposed a conceptual model based on Joiner's interpersonal theory of suicide, which proposes that low belonging and perceived burdensomeness are critical risk factors for suicide [141]. Based on this notion, Silva et al. suggest that cultural activities involving interactions with other Hispanics, such as those related to familism or religion, may increase engagement and feelings of belonging in Hispanic immigrant populations, in turn reducing suicide risk [142]. It is imperative to expand research on specific cultural interventions, given evidence that Hispanics are not a homogenous group. The differential suicide rates in Hispanic nationalities highlight the importance of the sociocultural context in understanding suicidality and the need for group-specific interventions (Tables 8.9 and 8.10).

Table 8.9 Key points: evidence-based interventions to prevent suicidality among Hispanic and Latino communities

- Compared to non-Hispanics, Hispanics are less likely to utilize mental health services. (see "barriers to accessing healthcare…" table).
- Interventions aimed at reducing the educational and language barrier gap among Hispanics may increase access to healthcare.
- Community-based interventions that increase cultural engagement have been found to decrease rate of suicidality in this population.

Table 8.10 Barriers to accessing healthcare among Latinos

Low educational achievement
Low income
High rates of uninsured migrants
Distrust of providers
Poor health literacy
Stigma towards mental health care
Language barrier

Discussion

Suicidal behavior in Hispanic and Latino populations varies widely by age, gender, ethnic subgroup, and experiences of immigration and acculturation. Certain risk factors are especially pertinent to Hispanic and Latino populations, such as young age, female gender, comorbid depression, low socioeconomic status, stressful life events, and greater acculturation. Immigration appears to play an important role in mental health outcomes and suicidality in Hispanic and Latino communities, as Hispanics who were born in or immigrated to the United States have higher rates of suicidal behavior than Hispanics who were not born in or did not immigrate to the United States. Certain risk factors related to the immigration process, including perceived discrimination and greater acculturation, evidenced in time spent in the new country and foreign language abilities, are critical to understanding risk for suicidality among Hispanics.

Although some Hispanic subgroups are at greater risk for suicidal behavior than non-Hispanics, research suggests that they may be less likely to die by suicide. Reconciling this so-called "Latino paradox" (Abraído-Lanza et al., 1999) involves understanding the strong protective factors against suicide among Hispanic and Latino populations. These protective factors, including familism, religiosity, and moral beliefs, may facilitate a stronger sense of ethnic or cultural belonging that safeguards against the alienation and isolation that can arise through acculturation.

Reducing suicidal behavior in Hispanic and Latino populations requires understanding the unique challenges and risk factors facing these communities. It is imperative to address the barriers to effective diagnosis and treatment of mental health issues in Hispanic communities, including lower educational attainment, socioeconomic status, health literacy, stigma, and lower English language proficiency. Impactful, lasting interventions must account for the diverse ethnic and cultural identities among Hispanic populations. They must be contextual, culturally specific, and embrace the protective factors already bolstering these communities.

References

1. Nock MK, Hwang I, Sampson NA, Kessler RC. Mental disorders, comorbidity and suicidal behavior: results from the National Comorbidity Survey Replication. Mol Psychiatry. 2010 Aug;15(8):868–76.
2. Baca-Garcia E, Perez-Rodriguez MM, Keyes KM, Oquendo MA, Hasin DS, Grant BF, et al. Suicidal ideation and suicide attempts in the United States: 1991-1992 and 2001-2002. Mol Psychiatry. 2010 Mar;15(3):250–9.
3. Turecki G, Brent DA. Suicide and suicidal behaviour. Lancet Lond Engl. 2016 Mar 19;387(10024):1227–39.
4. Silverman MM, Berman AL, Sanddal ND, O'carroll PW, Joiner TE. Rebuilding the tower of babel: a revised nomenclature for the study of suicide and suicidal behaviors. Part 2: suicide-related ideations, communications, and behaviors. Suicide Life Threat Behav. 2007 Jun;37(3):264–77.
5. Baca-Garcia E, Perez-Rodriguez MM, Oquendo MA, Keyes KM, Hasin DS, Grant BF, et al. Estimating risk for suicide attempt: are we asking the right questions? Passive suicidal ideation as a marker for suicidal behavior. J Affect Disord. 2011 Nov;134(1–3):327–32.
6. Fortuna LR, Álvarez K, Ramos Ortiz Z, Wang Y, Mozo Alegría X, Cook BL, et al. Mental health, migration stressors and suicidal ideation among Latino immigrants in Spain and the United States. Eur Psychiatry J Assoc Eur Psychiatr. 2016;36:15–22.
7. Nock MK, Borges G, Bromet EJ, Cha CB, Kessler RC, Lee S. Suicide and suicidal behavior. Epidemiol Rev. 2008;30:133–54.
8. Fortuna LR, Perez DJ, Canino G, Sribney W, Alegria M. Prevalence and correlates of lifetime suicidal ideation and suicide attempts among Latino subgroups in the United States. J Clin Psychiatry. 2007 Apr;68(4):572–81.
9. Perez-Rodriguez MM, Baca-Garcia E, Oquendo MA, Blanco C. Ethnic differences in suicidal ideation and attempts. Prim Psychiatry. 2008 Jan 1;15(2):44–53.
10. Abraído-Lanza AF, Dohrenwend BP, Ng-Mak DS, Turner JB. The Latino mortality paradox: a test of the "salmon bias" and healthy migrant hypotheses. Am J Public Health. 1999 Oct;89(10):1543–8.
11. Zayas LH, Pilat AM. Suicidal behavior in Latinas: explanatory cultural factors and implications for intervention. Suicide Life Threat Behav. 2008 Jun;38(3):334–42.
12. Romero AJ, Wiggs CB, Valencia C, Bauman S. Latina teen suicide and bullying. Hisp J Behav Sci. 2013 May 1;35(2):159–73.
13. Nock MK, Borges G, Bromet EJ, Alonso J, Angermeyer M, Beautrais A, et al. Cross-national prevalence and risk factors for suicidal ideation, plans and attempts. Br J Psychiatry J Ment Sci. 2008 Feb;192(2):98–105.
14. Balis T, Postolache TT. Ethnic differences in adolescent suicide in the United States. Int J Child Health Hum Dev IJCHD. 2008 Jan 1;1(3):281–96.
15. Kessler RC, Borges G, Walters EE. Prevalence of and risk factors for lifetime suicide attempts in the National Comorbidity Survey. Arch Gen Psychiatry. 1999 Jul;56(7):617–26.
16. Guiao IZ, Thompson EA. Ethnicity and problem behaviors among adolescent females in the United States. Health Care Women Int. 2004 Apr;25(4):296–310.
17. Grunbaum JA, Kann L, Kinchen SA, Ross JG, Gowda VR, Collins JL, et al. Youth risk behavior surveillance. National Alternative High School Youth Risk Behavior Survey, United States, 1998. J Sch Health. 2000 Jan;70(1):5–17.
18. Roberts RE, Roberts CR, Xing Y. Are Mexican American adolescents at greater risk of suicidal behaviors? Suicide Life Threat Behav. 2007 Feb;37(1):10–21.

19. Canino G, Roberts RE. Suicidal behavior among Latino youth. Suicide Life Threat Behav. 2001;31(Suppl):122–31.
20. Croyle KL. Self-harm experiences among Hispanic and non-Hispanic white young adults. Hisp J Behav Sci. 2007 May 1;29(2):242–53.
21. Grunbaum JA, Kann L, Kinchen SA, Williams B, Ross JG, Lowry R, et al. Youth risk behavior surveillance--United States, 2001. J Sch Health. 2002 Oct;72(8):313–28.
22. Klerman GL. Clinical epidemiology of suicide. J Clin Psychiatry. 1987 Dec;48(Suppl):33–8.
23. Spicer RS, Miller TR. Suicide acts in 8 states: incidence and case fatality rates by demographics and method. Am J Public Health. 2000 Dec;90(12):1885–91.
24. Mościcki EK, O'Carroll P, Rae DS, Locke BZ, Roy A, Regier DA. Suicide attempts in the epidemiologic catchment area study. Yale J Biol Med. 1988 Jun;61(3):259–68.
25. Suicide among racial/ethnic populations in the U.S.: American Indians/Alaska Natives [Internet]. Suicide Prevention Resource Center; 2011. Available from: https://www.samhsa.gov/node/669950
26. Kelder SH, Murray NG, Orpinas P, Prokhorov A, McReynolds L, Zhang Q, et al. Depression and substance use in minority middle-school students. Am J Public Health. 2001 May;91(5):761–6.
27. Schraedley PK, Gotlib IH, Hayward C. Gender differences in correlates of depressive symptoms in adolescents. J Adolesc Health Off Publ Soc Adolesc Med. 1999 Aug;25(2):98–108.
28. Eaton DK, Kann L, Kinchen S, Shanklin S, Ross J, Hawkins J, et al. Youth risk behavior surveillance–United States, 2007. Morb Mortal Wkly Rep Surveill Summ Wash DC 2002. 2008 Jun 6;57(4):1–131.
29. Eaton DK, Kann L, Kinchen S, Shanklin S, Flint KH, Hawkins J, et al. Youth risk behavior surveillance–United States, 2011. Morb Mortal Wkly Rep Surveill Summ Wash DC 2002. 2012 Jun 8;61(4):1–162.
30. Rasmussen KM, Negy C, Carlson R, Burns JM. Suicide ideation and acculturation among low socioeconomic status Mexican American adolescents. J Early Adolesc. 1997;17(4):390–407.
31. Roberts RE, Chen YW. Depressive symptoms and suicidal ideation among Mexican-origin and Anglo adolescents. J Am Acad Child Adolesc Psychiatry. 1995 Jan;34(1):81–90.
32. Hovey JD. Religion and suicidal ideation in a sample of Latin American immigrants. Psychol Rep. 1999 Aug;85(1):171–7.
33. Oquendo MA, Ellis SP, Greenwald S, Malone KM, Weissman MM, Mann JJ. Ethnic and sex differences in suicide rates relative to major depression in the United States. Am J Psychiatry. 2001 Oct;158(10):1652–8.
34. Sorenson SB, Golding JM. Suicide ideation and attempts in Hispanics and non-Hispanic whites: demographic and psychiatric disorder issues. Suicide Life Threat Behav. 1988;18(3):205–18.
35. Kessler RC, Berglund P, Borges G, Nock M, Wang PS. Trends in suicide ideation, plans, gestures, and attempts in the United States, 1990-1992 to 2001-2003. JAMA. 2005 May 25;293(20):2487–95.
36. Center for Behavioral Health Statistics and Quality. Results from the 2018 National Survey of Drug Use and Health (NSDUH): detailed tables [Internet]. Rockville, MD; 2019 [cited 2020 Jun 11]. Available from: https://www.samhsa.gov/data/report/2018-nsduh-detailed-tables
37. Monk M, Warshauer ME. Completed and attempted suicide in three ethnic groups. Am J Epidemiol. 1974 Oct;100(4):333–45.
38. Oquendo MA, Lizardi D, Greenwald S, Weissman MM, Mann JJ. Rates of lifetime suicide attempt and rates of lifetime major depression in different ethnic groups in the United States. Acta Psychiatr Scand. 2004 Dec;110(6):446–51.
39. Ungemack JA, Guarnaccia PJ. Suicidal ideation and suicide attempts among Mexican Americans, Puerto Ricans and Cuban Americans. Transcult Psychiatry. 1998 Jun 1;35(2):307–27.
40. Baca-Garcia E, Perez-Rodriguez MM, Keyes KM, Oquendo MA, Hasin DS, Grant BF, et al. Suicidal ideation and suicide attempts among Hispanic subgroups in the United States: 1991-1992 and 2001-2002. J Psychiatr Res. 2011 Apr;45(4):512–8.

41. Centers for Disease Control and Prevention (CDC), Youth risk behavior surveillance system. 1991-2017 High school youth risk behavior survey data [Internet] 2017. Available from: http://nccd.cdc.gov/youthonline/.
42. Rew L, Thomas N, Horner SD, Resnick MD, Beuhring T. Correlates of recent suicide attempts in a triethnic group of adolescents. J Nurs Scholarsh Off Publ Sigma Theta Tau Int Honor Soc Nurs. 2001;33(4):361–7.
43. Zayas LH, Lester RJ, Cabassa LJ, Fortuna LR. Why do so many Latina teens attempt suicide? A conceptual model for research. Am J Orthopsychiatry. 2005 Apr;75(2):275–87.
44. Vijayakumar L, John S, Pirkis J, Whiteford H. Suicide in developing countries (2): risk factors. Crisis. 2005;26(3):112–9.
45. Weissman MM, Bland RC, Canino GJ, Greenwald S, Hwu HG, Joyce PR, et al. Prevalence of suicide ideation and suicide attempts in nine countries. Psychol Med. 1999 Jan;29(1): 9–17.
46. Muehlenkamp JJ, Ertelt TW, Miller AL, Claes L. Borderline personality symptoms differentiate non-suicidal and suicidal self-injury in ethnically diverse adolescent outpatients. J Child Psychol Psychiatry. 2011 Feb;52(2):148–55.
47. Vega WA, Kolody B, Valle JR. Migration and mental health: an empirical test of depression risk factors among immigrant Mexican women. Int Migr Rev. 1987;21(3):512–30.
48. Brody DJ, Pratt LA, Hughes JP. Prevalence of depression among adults aged 20 and over: United States, 2013-2016. NCHS Data Brief. 2018;303:1–8.
49. Morales LS, Lara M, Kington RS, Valdez RO, Escarce JJ. Socioeconomic, cultural, and behavioral factors affecting Hispanic health outcomes. J Health Care Poor Underserved. 2002 Nov;13(4):477–503.
50. Hovey JD, King CA. Suicidality among acculturating Mexican Americans: current knowledge and directions for research. Suicide Life Threat Behav. 1997;27(1):92–103.
51. Alegria M, Canino G, Stinson FS, Grant BF. Nativity and DSM-IV psychiatric disorders among Puerto Ricans, Cuban Americans, and non-Latino whites in the United States: results from the National Epidemiologic Survey on alcohol and related conditions. J Clin Psychiatry. 2006 Jan;67(1):56–65.
52. Escobar JI, Hoyos Nervi C, Gara MA. Immigration and mental health: Mexican Americans in the United States. Harv Rev Psychiatry. 2000 Aug;8(2):64–72.
53. Perez-Rodriguez MM, Baca-Garcia E, Oquendo MA, Wang S, Wall MM, Liu S-M, et al. Relationship between acculturation, discrimination, and suicidal ideation and attempts among US Hispanics in the National Epidemiologic Survey of alcohol and related conditions. J Clin Psychiatry. 2014 Apr;75(4):399–407.
54. Oquendo MA, Dragatsi D, Harkavy-Friedman J, Dervic K, Currier D, Burke AK, et al. Protective factors against suicidal behavior in Latinos. J Nerv Ment Dis. 2005 Jul;193(7):438–43.
55. Kann L, Kinchen SA, Williams BI, Ross JG, Lowry R, Grunbaum JA, et al. Youth risk behavior surveillance--United States, 1999. MMWR CDC Surveill Summ Morb Mortal Wkly Rep CDC Surveill Summ. 2000 Jun 9;49(5):1–32.
56. Substance Abuse and Mental Health Services Administration. Summary of Findings from the 2000 National Household Survey on Drug Abuse | SAMHDA [Internet]. Rockville, MD; 2003 [cited 2020 Jun 12]. (DHHS Publication No. SMA 01–3549, NHSDA Series H-13). Available from: https://www.datafiles.samhsa.gov/study-publication/summary-findings-2000-national-household-survey-drug-abuse-nid14752.
57. De Oliveira G, Cianelli R, Gattamorta K, Kowalski N, Peragallo N. Social determinants of depression among Hispanic women. J Am Psychiatr Nurses Assoc. 2017 Feb;23(1): 28–36.
58. Kaminski JW, Puddy RW, Hall DM, Cashman SY, Crosby AE, Ortega LAG. The relative influence of different domains of social connectedness on self-directed violence in adolescence. J Youth Adolesc. 2010 May;39(5):460–73.
59. Arnett JJ. Emerging adulthood. A theory of development from the late teens through the twenties. Am Psychol. 2000 May;55(5):469–80.

60. Jacobson CM, Gould M. The epidemiology and phenomenology of non-suicidal self-injurious behavior among adolescents: a critical review of the literature. Arch Suicide Res Off J Int Acad Suicide Res. 2007;11(2):129–47.
61. Nock MK. Self-injury. Annu Rev Clin Psychol. 2010;6:339–63.
62. Nock MK, Joiner TE, Gordon KH, Lloyd-Richardson E, Prinstein MJ. Non-suicidal self-injury among adolescents: diagnostic correlates and relation to suicide attempts. Psychiatry Res. 2006 Sep 30;144(1):65–72.
63. Whitlock J, Eckenrode J, Silverman D. Self-injurious behaviors in a college population. Pediatrics. 2006 Jun;117(6):1939–48.
64. Wilcox HC, Arria AM, Caldeira KM, Vincent KB, Pinchevsky GM, O'Grady KE. Longitudinal predictors of past-year non-suicidal self-injury and motives among college students. Psychol Med. 2012 Apr;42(4):717–26.
65. Heath NL, Schaub K, Holly S, Nixon MK. Self-injury today: review of population and clinical studies in adolescents. In: Self-injury in youth: the essential guide to assessment and intervention. New York, NY, US: Routledge/Taylor & Francis Group; 2009. p. 9–27.
66. Sim L, Adrian M, Zeman J, Cassano M, Friedrich WN. Adolescent deliberate self-harm: linkages to emotion regulation and family emotional climate. J Res Adolesc. 2009;19(1):75–91.
67. Lloyd-Richardson EE, Perrine N, Dierker L, Kelley ML. Characteristics and functions of non-suicidal self-injury in a community sample of adolescents. Psychol Med. 2007 Aug;37(8):1183–92.
68. Guan K, Fox KR, Prinstein MJ. Nonsuicidal self-injury as a time-invariant predictor of adolescent suicide ideation and attempts in a diverse community sample. J Consult Clin Psychol. 2012 Oct;80(5):842–9.
69. Hankin BL, Abela JRZ. Nonsuicidal self-injury in adolescence: prospective rates and risk factors in a 2½ year longitudinal study. Psychiatry Res. 2011 Mar 30;186(1):65–70.
70. Hilt LM, Nock MK, Lloyd-Richardson EE, Prinstein MJ. Longitudinal study of nonsuicidal self-injury among young adolescents: rates, correlates, and preliminary test of an interpersonal model. J Early Adolesc [Internet] 2008 May 7 [cited 2020 Jun 12]; Available from: https://journals.sagepub.com/doi/10.1177/0272431608316604
71. Chesin MS, Moster AN, Jeglic EL. Non-suicidal self-injury among ethnically and racially diverse emerging adults: do factors unique to the minority experience matter? Curr Psychol. 2013 Dec 1;32(4):318–28.
72. Gratz KL. Risk factors for deliberate self-harm among female college students: the role and interaction of childhood maltreatment, emotional inexpressivity, and affect intensity/reactivity. Am J Orthopsychiatry. 2006 Apr;76(2):238–50.
73. Muehlenkamp JJ, Gutierrez PM. An investigation of differences between self-injurious behavior and suicide attempts in a sample of adolescents. Suicide Life Threat Behav. 2004;34(1):12–23.
74. Whitlock J, Muehlenkamp J, Eckenrode J. Variation in nonsuicidal self-injury: identification and features of latent classes in a college population of emerging adults. J Clin Child Adolesc Psychol Off J Soc Clin Child Adolesc Psychol Am Psychol Assoc Div 53. 2008 Oct;37(4):725–35.
75. Gratz KL, Conrad SD, Roemer L. Risk factors for deliberate self-harm among college students. Am J Orthopsychiatry. 2002 Jan;72(1):128–40.
76. Wester KL, Trepal HC. Nonsuicidal self-injury: exploring the connection among race, ethnic identity, and ethnic belonging. J Coll Stud Dev. 2015 Mar;56(2):127–39.
77. Baetens I, Claes L, Muehlenkamp J, Grietens H, Onghena P. Non-suicidal and suicidal self-injurious behavior among Flemish adolescents: a web-survey. Arch Suicide Res Off J Int Acad Suicide Res. 2011;15(1):56–67.
78. Baral I, Kora K, Yüksel Ş, Sezgin U. Self-mutilating behavior of sexually abused female adults in Turkey. J Interpers Violence. 1998;13(4):427–37.
79. Boudewyn AC, Liem JH. Childhood sexual abuse as a precursor to depression and self-destructive behavior in adulthood. J Trauma Stress. 1995 Jul;8(3):445–59.

80. Boxer P. Variations in risk and treatment factors among adolescents engaging in different types of deliberate self-harm in an inpatient sample. J Clin Child Adolesc Psychol Off J Soc Clin Child Adolesc Psychol Am Psychol Assoc Div 53. 2010;39(4):470–80.
81. Shapiro S. Self-mutilation and self-blame in incest victims. Am J Psychother. 1987 Jan;41(1):46–54.
82. van der Kolk BA, Perry JC, Herman JL. Childhood origins of self-destructive behavior. Am J Psychiatry. 1991 Dec;148(12):1665–71.
83. Wong JPS, Stewart SM, Ho SY, Lam TH. Risk factors associated with suicide attempts and other self-injury among Hong Kong adolescents. Suicide Life Threat Behav. 2007 Aug;37(4):453–66.
84. Donaldson D, Spirito A, Farnett E. The role of perfectionism and depressive cognitions in understanding the hopelessness experienced by adolescent suicide attempters. Child Psychiatry Hum Dev. 2000;31(2):99–111.
85. Goldston DB, Daniel SS, Reboussin BA, Reboussin DM, Frazier PH, Harris AE. Cognitive risk factors and suicide attempts among formerly hospitalized adolescents: a prospective naturalistic study. J Am Acad Child Adolesc Psychiatry. 2001 Jan;40(1):91–9.
86. Nock MK, Mendes WB. Physiological arousal, distress tolerance, and social problem-solving deficits among adolescent self-injurers. J Consult Clin Psychol. 2008 Feb;76(1):28–38.
87. van der Kolk BA. he complexity of adaptation to trauma: self-regulation, stimulus discrimination, and characterological development. In: Traumatic stress: the effects of overwhelming experience on mind, body, and society. New York, NY, US: Guilford Press; 1996. p. 182–213.
88. Gulbas LE, Hausmann-Stabile C, De Luca SM, Tyler TR, Zayas LH. An exploratory study of nonsuicidal self-injury and suicidal behaviors in adolescent Latinas. Am J Orthopsychiatry. 2015 Jul;85(4):302–14.
89. Brunner R, Kaess M, Parzer P, Fischer G, Carli V, Hoven CW, et al. Life-time prevalence and psychosocial correlates of adolescent direct self-injurious behavior: a comparative study of findings in 11 European countries. J Child Psychol Psychiatry. 2014 Apr;55(4):337–48.
90. Favazza AR, Conterio K. Female habitual self-mutilators. Acta Psychiatr Scand. 1989 Mar;79(3):283–9.
91. Walsh BW, Rosen PM. Self-mutilation: Theory, research, and treatment. New York, NY, US: Guilford Press; 1988. xiii, 273 p. (Self-mutilation: Theory, research, and treatment)
92. Zetterqvist M, Lundh L-G, Dahlström O, Svedin CG. Prevalence and function of non-suicidal self-injury (NSSI) in a community sample of adolescents, using suggested DSM-5 criteria for a potential NSSI disorder. J Abnorm Child Psychol. 2013 Jul;41(5):759–73.
93. American Psychiatric Association. Diagnostic and statistical manual of mental disorders [internet]. Fifth Edition American Psychiatric Association; 2013 [cited 2020 Feb 17]. Available from: https://psychiatryonline.org/doi/book/10.1176/appi.books.9780890425596.
94. Cervantes RC, Goldbach JT, Varela A, Santisteban DA. Self-harm among Hispanic adolescents: investigating the role of culture-related stressors. J Adolesc Health Off Publ Soc Adolesc Med. 2014 Nov;55(5):633–9.
95. Gholamrezaei M, De Stefano J, Heath NL. Nonsuicidal self-injury across cultures and ethnic and racial minorities: a review. Int J Psychol J Int Psychol. 2017 Aug;52(4):316–26.
96. Hovey JD. Acculturative stress, depression, and suicidal ideation in Mexican immigrants. Cultur Divers Ethnic Minor Psychol. 2000 May;6(2):134–51.
97. Vega WA, Gil AG, Zimmerman RS, Warheit GJ. Risk factors for suicidal behavior among Hispanic, African-American, and non-Hispanic white boys in early adolescence. Ethn Dis. 1993;3(3):229–41.
98. Burnam MA, Hough RL, Karno M, Escobar JI, Telles CA. Acculturation and lifetime prevalence of psychiatric disorders among Mexican Americans in Los Angeles. J Health Soc Behav. 1987 Mar;28(1):89–102.
99. Wadsworth T, Kubrin C. Hispanic Suicide in U.S. metropolitan areas: examining the effects of immigration, assimilation, affluence, and disadvantage [Internet]. Rochester, NY, Social Science Research Network; 2007. [cited 2020 Mar 24]. Report No.: ID 2028177. Available from: https://papers.ssrn.com/abstract=2028177.

100. Brown MJ, Cohen SA, Mezuk B. Duration of U.S. residence and suicidality among racial/ethnic minority immigrants. Soc Psychiatry Psychiatr Epidemiol. 2015 Feb;50(2):257–67.
101. Alderete E, Vega WA, Kolody B, Aguilar-Gaxiola S. Lifetime prevalence of and risk factors for psychiatric disorders among Mexican migrant farmworkers in California. Am J Public Health. 2000 Apr;90(4):608–14.
102. Sabogal F, Marín G, Otero-Sabogal R, Marín BV, Perez-Stable EJ. Hispanic Familism and acculturation: what changes and what Doesn't? Hisp J Behav Sci. 1987 Dec 1;9(4):397–412.
103. Gomez J, Miranda R, Polanco L. Acculturative stress, perceived discrimination, and vulnerability to suicide attempts among emerging adults. J Youth Adolesc. 2011 Nov;40(11):1465–76.
104. Oh H, Stickley A, Koyanagi A, Yau R, DeVylder JE. Discrimination and suicidality among racial and ethnic minorities in the United States. J Affect Disord. 2019;15(245):517–23.
105. Borges G, Breslau J, Su M, Miller M, Medina-Mora ME, Aguilar-Gaxiola S. Immigration and suicidal behavior among Mexicans and Mexican Americans. Am J Public Health. 2009 Apr;99(4):728–33.
106. Barranco RE. Suicide, Religion, and Latinos: a macrolevel study of U.S. Latino suicide rates. Sociol Q. 2016;57(2):256–81.
107. Singh GK, Hiatt RA. Trends and disparities in socioeconomic and behavioural characteristics, life expectancy, and cause-specific mortality of native-born and foreign-born populations in the United States, 1979-2003. Int J Epidemiol. 2006 Aug;35(4):903–19.
108. Peña JB, Wyman PA, Brown CH, Matthieu MM, Olivares TE, Hartel D, et al. Immigration generation status and its association with suicide attempts, substance use, and depressive symptoms among latino adolescents in the USA. Prev Sci Off J Soc Prev Res. 2008 Dec;9(4):299–310.
109. Alegria M, Shrout PE, Woo M, Guarnaccia P, Sribney W, Vila D, et al. Understanding differences in past year psychiatric disorders for Latinos living in the US. Soc Sci Med 1982. 2007 Jul;65(2):214–30.
110. Guarnaccia PJ, Pincay IM, Alegria M, Shrout P, Lewis-Fernandez R, Canino G. Assessing diversity among Latinos: results from the NLAAS. Hisp J Behav Sci. 2007;29(4):510–34.
111. Walker RL, Wingate LR, Obasi EM, Joiner TE. An empirical investigation of acculturative stress and ethnic identity as moderators for depression and suicidal ideation in college students. Cultur Divers Ethnic Minor Psychol. 2008 Jan;14(1):75–82.
112. Borowsky IW, Ireland M, Resnick MD. Adolescent suicide attempts: risks and protectors. Pediatrics. 2001 Mar;107(3):485–93.
113. Locke TF, Newcomb MD. Psychosocial predictors and correlates of suicidality in teenage Latino males. Hisp J Behav Sci. 2005;27(3):319–36.
114. Zayas LH, Kaplan C, Turner S, Romano K, Gonzalez-Ramos G. Understanding suicide attempts by adolescent Hispanic females. Soc Work. 2000 Jan 1;45(1):53–63.
115. NW 1615 L. St, Suite 800 Washington, Inquiries D 20036USA202-419-4300 | M-857-8562 | F-419-4372 | M. When Labels Don't fit: Hispanics and their views of identity [internet]. Pew Research Center's Hispanic Trends Project 2012 [cited 2020 Apr 6]. Available from: https://www.pewresearch.org/hispanic/2012/04/04/when-labels-dont-fit-hispanics-and-their-views-of-identity/.
116. Robinson JA, Bolton JM, Rasic D, Sareen J. Exploring the relationship between religious service attendance, mental disorders, and suicidality among different ethnic groups: results from a nationally representative survey. Depress Anxiety. 2012 Nov;29(11):983–90.
117. Pescosolido BA, Georgianna S. Durkheim, suicide, and religion: toward a network theory of suicide. Am Sociol Rev. 1989;54(1):33–48.
118. Trovato F, Jarvis GK. Immigrant suicide in Canada: 1971 and 1981. Soc Forces. 1986;65(2):433–57.
119. van Tubergen F, te Grotenhuis M, Ultee W. Denomination, religious context, and suicide: neo-durkheimian multilevel explanations tested with individual and contextual data. Am J Sociol. 2005;111(3):797–823.
120. Suicide: A Study in Sociology. Emile Durkheim, John A. Spaulding, George Simpson. Am J Sociol. 1951 Jul 1;57(1):100–1.

121. Ahmedani BK, Perron B, Ilgen M, Abdon A, Vaughn M, Epperson M. Suicide thoughts and attempts and psychiatric treatment utilization: informing prevention strategies. Psychiatr Serv Wash DC. 2012 Feb 1;63(2):186–9.
122. Badawi M, Kramer M, Eaton W. Use of mental health services by households in the United States. Psychiatr Serv Wash DC. 1996 Apr;47(4):376–80.
123. Cabassa LJ, Zayas LH, Hansen MC. Latino adults' access to mental health care: a review of epidemiological studies. Admin Pol Ment Health. 2006 May;33(3):316–30.
124. Keyes KM, Martins SS, Hatzenbuehler ML, Blanco C, Bates LM, Hasin DS. Mental health service utilization for psychiatric disorders among Latinos living in the United States: the role of ethnic subgroup, ethnic identity, and language/social preferences. Soc Psychiatry Psychiatr Epidemiol. 2012 Mar;47(3):383–94.
125. Wells KB, Golding JM, Hough RL, Burnam MA, Karno M. Acculturation and the probability of use of health services by Mexican Americans. Health Serv Res. 1989 Jun;24(2): 237–57.
126. Alegría M, Canino G, Ríos R, Vera M, Calderón J, Rusch D, et al. Inequalities in use of specialty mental health services among Latinos, African Americans, and non-Latino whites. Psychiatr Serv Wash DC. 2002 Dec;53(12):1547–55.
127. Ginzberg E. Access to health care for Hispanics. JAMA. 1991 Jan 9;265(2):238–41.
128. Treviño FM, Moyer ME, Valdez RB, Stroup-Benham CA. Health insurance coverage and utilization of health services by Mexican Americans, mainland Puerto Ricans, and Cuban Americans. JAMA. 1991 Jan 9;265(2):233–7.
129. Raymond-Flesch M, Siemons R, Pourat N, Jacobs K, Brindis CD. There is no help out there and if there is, it's really hard to find: a qualitative study of the health concerns and health care access of Latino "dreamers". J Adolesc Health Off Publ Soc Adolesc Med. 2014 Sep;55(3):323–8.
130. Ortega AN, McKenna RM, Kemmick Pintor J, Langellier BA, Roby DH, Pourat N, et al. Health care access and physical and behavioral health among undocumented Latinos in California. Med Care. 2018 Nov;56(11):919–26.
131. de Diaz-Perez M, Farley T, Cabanis CM. A program to improve access to health care among Mexican immigrants in rural Colorado. J Rural Health Off J Am Rural Health Assoc Natl Rural Health Care Assoc. 2004;20(3):258–64.
132. Obando Medina C, Kullgren G, Dahlblom K. A qualitative study on primary health care professionals' perceptions of mental health, suicidal problems and help-seeking among young people in Nicaragua. BMC Fam Pract. 2014 Jul;2(15):129.
133. Rotheram-Borus MJ, Piacentini J, Van Rossem R, Graae F, Cantwell C, Castro-Blanco D, et al. Treatment adherence among Latina female adolescent suicide attempters. Suicide Life Threat Behav. 1999;29(4):319–31.
134. Dueweke AR, Bridges AJ. The effects of brief, passive psychoeducation on suicide literacy, stigma, and attitudes toward help-seeking among Latino immigrants living in the United States. 2017;
135. Timmins CL. The impact of language barriers on the health care of Latinos in the United States: a review of the literature and guidelines for practice. J Midwifery Womens Health. 2002 Apr;47(2):80–96.
136. Ortega P. Spanish language concordance in U.S. medical care: a multifaceted challenge and call to action. Acad Med J Assoc Am Med Coll. 2018;93(9):1276–80.
137. Perrino T, Pantin H, Prado G, Huang S, Brincks A, Howe G, et al. Preventing internalizing symptoms among Hispanic adolescents: a synthesis across Familias Unidas trials. Prev Sci Off J Soc Prev Res. 2014 Dec;15(6):917–28.
138. Brincks A, Perrino T, Howe G, Pantin H, Prado G, Huang S, et al. Preventing youth internalizing symptoms through the Familias Unidas intervention: examining variation in response. Prev Sci Off J Soc Prev Res. 2018;19(Suppl 1):49–59.
139. Vidot DC, Huang S, Poma S, Estrada Y, Lee TK, Prado G. Familias Unidas' crossover effects on suicidal behaviors among Hispanic adolescents: results from an effectiveness trial. Suicide Life Threat Behav. 2016 Apr;46(Suppl 1):S8–14.

140. Hall M, Fullerton L, FitzGerald C, Green D. Suicide risk and resiliency factors among Hispanic teens in New Mexico: schools can make a difference. J Sch Health. 2018;88(3):227–36.
141. Van Orden KA, Witte TK, Cukrowicz KC, Braithwaite S, Selby EA, Joiner TE. The interpersonal theory of suicide. Psychol Rev. 2010 Apr;117(2):575–600.
142. Silva C, Van Orden KA. Suicide among Hispanics in the United States. Curr Opin Psychol. 2018 Aug;22:44–9.

9 Substance Use in the Latino Population in the United States

Youssef Mahfoud, Michael Ignatowski, and Jessica Gonzalez-Tindle

Substance use disorders involve the recurrent use of alcohol and/or other drugs leading to consumption of amounts larger than intended, use in hazardous situations, and the inability to stop using a substance. The use leads to clinical and functional impairment, which can include health problems, disability, or social deficiency such as failure to meet major work, school, or home responsibilities. Physical addiction such as tolerance and withdrawal symptoms often develop with continued use. The substance use disorders can be defined as mild, moderate, and severe depending on the degree of impairment and the number of diagnostic criteria met [1].

The 2018 National Survey on Drug Use and Health (NSDUH) in the United States estimates that over 55% of adults used alcohol in the past month. Another 9.9 million reported misusing prescription pain relievers, and over 808,000 reported using heroin in the past year. Overall, 7.8% of the adult population (19.3 million) likely meet criteria for a substance use disorder. Among those with a substance use disorder, 74.5% (14.4 million) struggled with alcohol use, 38.% (7.4 million) struggled with illicit drug use, and 12.9% (2.5 million) struggled with both illicit drugs and alcohol [2].

Estimates of the rate of substance use in the Latino population in the United States have emerged over the past decades with collection of NSDUH ethnicity data. The 2018 survey revealed that 7.6% of Latino adults (3.1 million) likely meet criteria for a substance use disorder, only slightly less than the overall adult population. Among those with substance use disorders, 77.1% (2.4 million) struggle with alcohol use, 39.5% struggle with illicit drug use (1.2 million), and 16.6% struggle with both [2].

Y. Mahfoud (✉) · M. Ignatowski
Northeast Ohio VA medical Center, Cleveland, OH, USA

Case Western Reserve University School of Medicine, Cleveland, OH, USA
e-mail: Youssef.Mahfoud@va.gov; Michael.Ignatowski2@va.gov

J. Gonzalez-Tindle
Northeast Ohio VA medical Center, Cleveland, OH, USA
e-mail: Jessica.Tindle@va.gov

R. Castilla-Puentes, T. Falcone (eds.), *Mental Health for Hispanic Communities*,
https://doi.org/10.1007/978-3-031-13195-0_9

Latinos report using less alcohol in the past month compared to the national average in all age categories: ages 12–17 (7.4% vs 9%), ages 18–25 (48.9% vs. 55.1%), and 26+ (46.4% vs. 55.3%). Despite this, rates of alcohol dependence (5.3%) and binge drinking (24.6%) among Hispanics are similar to those of European Americans and slightly higher than those of African Americans [2].

Although past year and past month drug use is similar to other ethnic groups, Hispanic Americans report lower rates of lifetime illicit drug use (37.7%) than European (54.5%) and African Americans (45.9%). The exception to this is ages 12–17, lifetime adolescent use rates increase more than non-Hispanic White Americans (25.2% vs. 23.4%).

Opioid misuse was reported by 3.6% of Hispanic Americans in the past year which is similar to the general population, although the use of heroin among those who misuse is less in the Hispanic American population (6.4% vs 7.9%). Heroin-related opioid use disorders were also slightly lower than the general population (0.1% vs. 0.3%) [2]. Occupational exposures may shift the Hispanic population to prescription opioid pain pill use rather than heroin. In 2017 the overdose death rate among Hispanic Americans was 6.8 deaths per 100,000, significantly lower than the national average of 14.9 deaths per 100,000 [3].

Rates of adolescent marijuana use in the past month are similar to the general population, but rates of adult marijuana use are reported less than the general population, in ages 18–25 (19.3% vs. 22.1%), and ages 26 and older (6.5% vs. 8.6%). Rates of marijuana use disorder are less than the general population for ages 18–25 (4.8% vs. 5.9%), but similar in the 26 or older category (both 0.9%). Cocaine and methamphetamine use is similar to the national average.

As of 2018, prevalence of tobacco use among Hispanics is lower than the general population (9.8% vs. 13.7%) [4]. Hispanics have lower health insurance coverage and less healthcare access than non-Hispanic Whites translating into less counselling and less access to cessation treatments. Nonetheless, they quit smoking at higher rates than non-Hispanic Whites and the general population [5].

Role of Acculturation

Studies suggest that there is likely to be an increasing prevalence of substance use disorders among Hispanics that may be attributable to increasing levels of assimilation to dominant culture in the United States [6].

A positive correlation exists between substance use and proficiency and preference with the English language, compared to those primarily speaking Spanish [7, 8]. Those that have undergone acculturation in the United States may have use rates more similar to the national average. Assimilated language may represent a loss of closeness with family, and stronger connection to peers. One consideration, however, is that individuals who predominantly speak Spanish may be more reluctant to report on national surveys [9].

National survey data indicates that Hispanic individuals immigrating to the United States have a lower chance of using alcohol and drugs than those born in the

United States [10]. Those born in the USA or residing for over 15 years, had higher rates of alcohol use. Similar results are seen with tobacco use. The prevalence of cigarette smoking is higher among Hispanic adults born in the United States than those who immigrated [11]. Having one parent born in the United States also appears to be a risk factor for developing a substance use disorder [6].

Gender and Substance Use

Rates of substance use among Hispanic females have historically been lower than males [12]. Rates of use in Hispanic women tend to be lower than non-Hispanic White women as well [13]. Most significant is the NSDUH survey data related to pregnancy. Rates of past month substance use among pregnant Hispanic women are significantly lower than the general population. The categories showing this trend include alcohol use (6.2% vs 9.9%), tobacco use (5.2% vs. 11.6%), marijuana use (1% vs. 4.7%), and overall illicit drug use (1% vs. 5.4%). Hispanic women generally have a low prevalence of cigarette smoking during pregnancy as well. Most significant is the NSDUH survey data related to pregnancy. Rates of past month substance use among pregnant Hispanic women are significantly lower than the general population. The categories showing this trend include alcohol use (6.2% vs 9.9%), tobacco use (5.2% vs. 11.6%), marijuana use (1% vs. 4.7%), and overall illicit drug use (1% vs. 5.4%) [2]. Hispanic women generally have a low prevalence of cigarette smoking during pregnancy as well [14].

Differences in Latin American Place of Origin

Hispanic Americans are often studied as a single population, but important differences arise when looking at Hispanic subgroups. Puerto Ricans have the highest rate of recent illicit drug use including marijuana and South Americans have the lowest [15].

A 2014 study indicated that Puerto Rican men and women tend to be the heaviest drinkers as well compared to other Hispanic Americans. Puerto Rican Americans have a threefold chance of developing an alcohol use disorder than non-Hispanic White Americans. This may be due in part to the acceptance of earlier alcohol initiation and lack of stigma compared to other regions. The study also notes that Cuban men drink the least of Hispanic men and Mexican women drink the least of Hispanic women. Beer is the preferred drink across all Hispanic subgroups in the United States [16].

Cigarette smoking prevalence varies by Hispanic/Latino subgroups, as well, tending to also be higher among the Puerto Rican community [17]. Among daily smokers, Cuban daily smokers tend to smoke more cigarettes per day, while those of Mexican heritage tend to be the least likely to report smoking 20 cigarettes per day or more, and are the most likely group to be intermittent smokers as opposed to daily, 15.5% compared to 9.8% of Central American men, 9% of Puerto Rican men, and 4.9% of Cuban men [18].

Treatment

Despite the wealth of information regarding the prevalence of substance use in the Latino population, there is a paucity of research assessing the effectiveness of treatment modalities in this population. Of the 3.3 million Latinos with substance use disorders, 89.7% received no treatment, which is similar to the national average. Of note, the 1.3 million with dual diagnoses of mental illness and substance use are more likely to get no treatment than the general population (93% vs. 90.4%). In the overall adult population, Hispanic Americans are less likely to get specialty substance use care than the general population (0.6% vs. 1%) [2]. Compared to those in treatment, individuals with no treatment are more likely to have poor outcomes that can include development of chronic health conditions, Hepatitis C, HIV, premature mortality, and interactions with the criminal justice system [19].

Other national surveys also have shown that Latinos have less access to specialized substance use treatment than the rest of the population, their wait time to access services is longer and they are less satisfied with the treatment [20–22]. Studies have also shown that Latinos were less likely to express a need for treatment and to participate in it than other ethnicities [23, 24] and that utilization of treatment was even lower in Latino immigrants than in those born in the USA [25, 26]. Shame and concerns about disrupting family relationships may be the reasons why some Latinos avoid checking in into residential treatment according to some studies [27, 28].

Additional studies have found that Latinos are more likely to terminate prematurely outpatient substance use treatment [29–31] and more than half of this population using mental health services discontinue therapy after only one session [32]. Some of the reasons noted for premature termination are the lack of bilingual and bicultural therapist and failed outreach, recruitment, and retention strategies [33, 34]. However, in contrast to the findings that showed poorer outcomes and premature treatment termination, some studies involving alcohol use have found similar outcomes to those of other ethnic groups [35, 36].

The cultural mismatch between patients and clinicians may explain some of studies that revealed a higher early termination rate and the poorer outcomes of Latinos in substance use services. The importance of considering culture in delivering substance use treatment has been pointed out in a certain number of studies [37–39]. Latinos endorsed the positive impact of having Latino providers as well as the importance of family and religion and spirituality in recovery.

Cultural Aspects to Consider When Treating Substance Use in the Latino Population

It is crucial that clinicians take into account the cultural framework and select culturally informed interventions. Culturally informed treatment interventions are: (1) *based on the cultural values of the group,* (2) *consider the attitudes, expectations, and norms of the group, and* (3) *consider behavioral preferences and expectations of group members* [40].

It is important to note that substance use treatment is not culturally accepted in the Latino population. While Latinos have significantly higher drug addiction stigma scores than that of African Americans, stigma has been shown to decrease with acculturation [41]. Although there remains a gap in the literature to understand the barriers behind Latino's low utilization of substance use treatment, especially those with co-occurring mental illness [2, 42], one qualitative study has found that Latinos largely avoided specialty treatment due to: (1) beliefs that treatment is not culturally informed, (2) perceptions that substance use treatment was ineffective, (3) a fear of judgment from providers related to differing recovery goals (e.g., if a patient was not seeking abstinence), (4) a lower perceived need for treatment (if able to maintain familial duties), (5) a lack of support from their family, and last but certainly not least (6) stigma, which contributed to frequent descriptions of seeing treatment as a "personal failure" or "defeat" [41].

Familismo (familism) is a term used in the Latino culture to identify the strong influence of the family system on an individual; there is a strong emphasis on family dynamics, extended social networks, and the collective resources shared and provided within the family system [43]. Furthermore, familismo is characterized by a strong sense of loyalty, reciprocity, and solidarity [44]. There is also a strong sense of duty to protect the emotional and physical well-being of the family unit [45], which may have the potential to inadvertently exacerbate substance use. If a Latino client feels that a therapist is soliciting disclosure of information that may bring about shame to the family, they may minimize or withhold information, and may ultimately withdraw from services. It will be essential to not imply blame towards family system as a means for continued substance use, but to influence the family system to support non-use, instead [46]. Therapy that stresses role and responsibilities of each family member should be considered. For example, re-establishing family ties that have been severed due to substance use [47].

Simpatía can be defined as the *value of behaviors that contribute to smooth and pleasant interactions* [46]. In an earnest attempt to establish or maintain the sense of harmony and respect congruent with this value, there is a general tendency for Latinos to avoid interpersonal conflict, minimize the impact of negative behaviors, and emphasize agreement in social situations [48, 49]. In the context of a Latino patient, with a substance use disorder, the value of minimizing conflict may result in the patient's family system unintentionally enabling continued substance use, and may shield the user from natural consequences associated with their use. In this example, it is recommend the provider explore how continued avoidance of addressing behaviors associated with substance use, promotes (rather than prevent) future conflict.

In the Latino population, personalismo is a style of communication, in which there is a preference for personal interactions over impersonal [50]. Developing an authentic relationship with a Latino client, as they are inviting you into their family system, is especially important. It is not uncommon for a Latino client to ask a therapist about their personal experiences, in an attempt to develop an interpersonal connection and to develop trust. Failure to establish a personal connection with a Latino client may increase the odds of the client dropping out of treatment prematurely [46].

Machismo is defined as a socio-cultural construct that describes the beliefs and expectations of masculinity [51]. The construct of Latino families is largely influenced by a patriarchal structure and traditional gender roles, where males are expected to be strong and provide to the family [46]. Common characteristics of machismo include: bravery, honor, dominance, aggression, sexual prowess, reserved emotions [52, 53]. While there is limited literature related to masculinity and addiction, Davis [54] describes heavy drinking as being socially sanctioned, whereas the mere labeling of a Latino male as an addict elicits feelings of shame and moral weakness. The Latino culture defines alcoholism as an individual's inability to uphold their social responsibilities, undermining their sense of masculinity (machismo). It will be important to explore the extent the Latino client's internalizing beliefs, related to being in substance use treatment. Culturally informed interventions should then be considered subsequently.

Marianismo is the inverse construct of machismo, that describes the beliefs and expectations of female gender roles [51]. In the Latino culture, women are expected to be nurturing, submissive, and self-sacrificing to the male [55–58]. Other common characteristics of marianismo include: family and home centered, passive, and chastity [53, 59]. Boyd-Franklin and Garcia Preto [55] have described Latina women as living in a "cultural paradox," due to the contradictory expectation that they are seen as "morally and spiritually superior to men", yet are expected to accept male authority. Contrary to machismo, it is thought that the cultural expectations of marianismo serve as a protective factor for a Latina woman to develop substance use disorders, by encouraging abstinence and may be especially protective during pregnancy compared to other ethnicities [60–62].

As many as 82% of Latinos identify with a religion [63]. The value of spiritualism in the Latino Culture is largely influenced by Roman Catholic practices, as Catholism is the predominate religion in the Latino population. Within this religious practice is the belief of spirits and divine interventions [50]. There may be a belief that the problem of substance use/abuse is attributable to God's will, witchcraft [64, 65], or as the influence of some higher order spiritual power [46]. When working with a Latino client, it will be helpful to explore their belief system, regarding spiritual factors that they feel sustain their substance use [64].

In summary, although Latinos have statistically lower rates of substance use disorders, these rates have been shown to increase with US acculturation. Furthermore, Latinos have less access to treatment than other minority groups and lower rates of continuation of services than the general population. It is recommended that clinicians incorporate culturally informed interventions, when working with Latino clients, to increase access to services, reduce language barriers, and overall improve treatment outcomes for Latinos with substance use disorders.

There is definitely a need for more research about the impact of various types of substances use treatments and interventions on the Latino population as well as the effect of incorporating culturally sensitive protocols in the treatment plan.

References

1. American Psychiatric Association. Diagnostic and statistical manual of mental disorders (DSM-5). 5th ed. Arlington, VA: American Psychiatric Publishing; 2013.
2. Center for Behavioral Health Statistics and Quality. Results from the 2018 National survey on drug use and health: detailed tables. Rockville, MD: Substance Abuse and Mental Health Services Administration; 2019. Accessed 2020 Oct 7.
3. Centers for Disease Control and Prevention. 2019 Annual surveillance report of drug-related risks and outcomes—United States. [internet]. Atlanta, GA: CDC National Center for Injury Prevention and Control.; Accessed 2020 Oct 7.
4. Creamer MR, Wang TW, Babb S, et al. Tobacco Product Use and Cessation Indicators Among Adults–United States, 2018. Morb Mortal Wkly Rep. 2019;68(45):1013–9.
5. U.S. Department of Health and Human Services. Tobacco Use Among U.S. Racial/Ethnic Minority Groups–African Americans, American Indians and Alaska Natives, Asian Americans and Pacific Islanders, Hispanics: a report of the surgeon general. Atlanta: U.S. Department of Health and Human Services, Centers for Disease Control and Prevention, National Center for Chronic Disease Prevention and Health Promotion, Office on Smoking and Health; 1998. Accessed 2020 Oct 7.
6. Ortega AN, Rosenheck R, Alegría M, Desai RA. Acculturation and the lifetime risk of psychiatric and substance use disorders among Hispanics. J Nerv Ment Dis. 2000;188(11):728–35.
7. Black SA, Markides K. Acculturation and alcohol consumption in Puerto Rican, Cuban American, and Mexican-American women in the United States. Am J Public Health. 1993;83:890–3.
8. Finch BK. Nation of origin, gender, and neighborhood differences in past year substance use among Hispanics and non-Hispanic whites. Hisp J Behav Sci. 2001;23:88–101.
9. Wagner-Echeagaray F, Chilcoat HD, Anthony JC. Degree of acculturation and the risk of crack cocaine smoking among Hispanic Americans. Br J Psychiatry. 1994;168:17–30.
10. Abraido-Lanza A, Chao M, Flórez K. Do healthy behaviors decline with greater acculturation? Implications for the Latino mortality paradox. Soc Sci Med. 2005;61:1243–55. https://doi.org/10.1016/j.socscimed.2005.01.016.
11. SAMSHA. Substance abuse and mental health services administration. Results from the 2016 national survey on drug use and health: detailed tables pdf icon external icon. Rockville, MD: Substance Abuse and Mental Health Services Administration, Center for Behavioral Health Statistics and Quality; 2016. Accessed 2020 Oct 7.
12. Anthony JC, Warner LA, Kessler RC. Comparative epidemiology of dependence on tobacco, alcohol, controlled substances, and inhalants: basic findings from the National Comorbidity Survey. Exp Clin Psychopharmacol. 1994;2(3):244–68.
13. Young VD, Harrison R. Race/ethnic differences in the sequences of drugs used by women. J Drug Issues. 2001;31(2):293–324.
14. Centers for Disease Control and Prevention. Trends in smoking before, during, and after pregnancy–pregnancy risk assessment monitoring system, United States, 40 Sites, 2000–2010. Morb Mortal Wkly Rep. 2013;62(SS06):1–19. Accessed 2020 Oct 7.
15. National Institute on Drug Abuse. Drug use among racial/ethnic minorities. 2003.
16. Ríos-Bedoya CF, Freile-Salinas D. Incidence of alcohol use disorders among hispanic subgroups in the USA. Alcohol Alcohol. 2014;49(5):549–56.
17. CDC Centers for Disease Control and Prevention. Disparities in adult cigarette smoking–United States, 2002-2005 and 2010-2013. Morb Mortal Wkly Rep. 2016;65(30). Accessed 2020 Oct 7.
18. Kaplan RC, Bangdiwala SI, Barnhart JM. Smoking among U.S. Hispanic/Latino adults: the Hispanic community health study/study of Latinos. Am J Prev Med. 2014;46(5):496–506.

19. National Drug Intelligence Center. National drug threat assessment. Washington, DC: U.S. Department of Justice; 2011.
20. Substance Abuse and Mental Health Services Administration. National household survey on drug abuse. 2001.
21. Substance Abuse and Mental Health Services Administration. National survey on drug use and health. 2002.
22. Wells K, Klap R, Koike A, Sherbourne C. Ethnic disparities in unmet need for alcoholism, drug abuse and mental health care. Am J Psychiatr. 2001;158:2027–32.
23. Longshore D, Hsieh S, Anglin MD. Ethnic and gender differences in drug user's perceived need for treatment. Int J Addict. 1993;28:539–58.
24. Anglin MD, Hser YI, Grella CE. Drug addiction and treatment careers among clients in the drug abuse treatment outcome study (DATOS). Psychol Addict Behav. 1997;11:308–11.
25. Fisher DG, Reynolds GL, Moreno-Branson CM, Jaffe A, Wood MM, Klahn JA, Muniz JF. Drug treatment needs of Hispanic drug users in Long Beach. CA Journal of Drug Issues. 2004;34:879–94.
26. Vega WA, Sribney WM. Seeking care for alcohol problems: patterns of need and treatment among Mexican-origin adults in Central California. Alcohol Treat Q. 2005;23:29–51.
27. Amaro H, Nieves R, Johannes SW, Cabeza NML. Substance abuse treatment: critical issues and challenges in the treatment of Latina women. Hisp J Behav Sci. 1999;21:266–82.
28. Kline A. Pathways into drug user treatment: the influence of gender and racial/ethnic identity. Subst Use Misuse. 1996;31:323–42.
29. Agosti V, Nunes E, Ocepeck-Welikson K. Patient factors related to early attrition from outpatient cocaine research. Am J Drug Alcohol Abuse. 1996;22:29–41.
30. Hser YI, Huang D, Teruya C, Anglin MD. Diversity of drug abuse treatment utilization patterns and outcomes. Eval Program Plann. 2005;27:309–19.
31. Kleinman PH, Kang SY, Lipton DS, Woody GE, Kemp J, Millman RB. Retention of cocaine abusers in outpatient psychotherapy. Am J Drug Alcohol Abuse. 1992;18:29–43.
32. Gloria AM, Peregoy JJ. Counseling Latino alcohol and other substance users/abusers: cultural considerations for counselors. J Subst Abus Treat. 1996;13(2):119–26.
33. Glick R, Moore JE. Drugs in Hispanic communities. New Brunswick, NJ: Rutgers University Press; 1990.
34. SAMHSA. Drug Abuse warning network. Washington, DC: Office of Applied Studies; 2002.
35. Bell DC, Williams ML, Nelson R, Spence RT. An experimental test of retention in residential and outpatient programs. Am J Drug Alcohol Abuse. 1994;20:331–40.
36. Mulvaney FD, et al. Methadone maintenance outcomes for Hispanic and African American men and women. Drug Alcohol Depend. 1999;54:11–8.
37. Trepper T, Nelson TS, McCollum EE, McAvoy P. Improving substance abuse service delivery to Hispanic women through increased cultural competencies: a qualitative study. J Subst Abus Treat. 1997;14:225–34.
38. Hohman M. Treatment experiences of women in a recovery home for Latinas. Alcohol Treat Q. 1999;17:67–78.
39. Porter J. The street/treatment barrier: treatment experiences of Puerto Rican injection drug users. Subst Use Misuse. 1999;34:1951–75.
40. De La Rosa M, White M. The role of social support systems in the drug use behavior of Hispanics revisited. J Psychoactive Drugs. 2001;33:233–41.
41. Pinedo M, Zemore S, Rogers S. Understanding barriers to specialty substance abuse treatment among Latinos. J Subst Abus Treat. 2018;94:1–8.
42. Guerrero EG, Marsh JC, Khachikian T, Amaro H, Vega WA. Disparities in Latino substance use, service use, and treatment: Implications for culturally and evidence-based interventions under health care reform. Drug and Alcohol Dependence. 2013;133(3):805–13.
43. Lopez-Tamayo R, Seda A, Jason LA. The role of familismo and acculturation as moderators of the association between family conflict and substance abuse on Latino adult males. Public health (Fairfax) [Internet]. 2016 Aug [cited 2020 Oct 7];1(2):48. Available from: https://www.ncbi.nlm.nih.gov/pmc/articles/PMC5354469/.

44. Marín G, Marín BV. Research with Hispanic populations. Newbury Park, CA: Sage Publishers, Inc; 1991.
45. Cohen R. Culture, disease and stress among Latino immigrants. Washington, DC: Smithsonian Institution; 1979.
46. Torres-Rivera E, Wilbur MP, Phan LT, Maddux CD, Roberts-Wilbur J. Counseling Latinos with substance Abuse problems. J Addict Offender Couns. 2011;13(2):26–42.
47. Gilbert MJ, Cervantes RC. Alcohol services for Mexican- Americans: a review of utilization patterns, treatment considerations, and prevention activities. Hisp J Behav Sci. 1986;8: 191–223.
48. Kagan S, Knight GP, Martinez-Romero S. Culture and the development of conflict resolution style. J Cross-Cultural Psychol. 1982;13(1):43–59. Kimmel, M. (2008)
49. Trandis HC, Marín G, Betancourt H. Simpatía as a cultural script of Hispanics. J Pers Soc Psychol. 1984;47(6):1363–75.
50. Comas-Díaz L. An integrative approach. In: Comas-Díaz L, Greene B, editors. Women of color: integrating ethnic and gender identities in psychotherapy. The Guilford Press; 1994. p. 287–318.
51. Nuñez A, Gonzalez P, Talavera GA, Roesch SC, Arguelles W, Ostrovsky NW, Penedo FJ, Sanchez-Johnsen L, Davis SM, Womack VY, Ojeda L, Gallo LC. Machismo, Marianismo, and negative cognitive- emotional factors: findings from the Hispanic community health study/ study of Latinos sociocultural ancillary study. Journal of Latina/o Psychology. 2016;4(4): 202–17.
52. Mirandé A. Hombres y machos: masculinity and Latino culture. Boulder, CO: Westview Press; 1977.
53. Niemann YF. Stereotypes of Chicanas and Chicanos: impact on family functioning, individual expectations, goals, and behavior. In: Velasquez RJ, Arellano LM, McNell BW, editors. The handbook of Chicana/o psychology and mental health. Mahwah, NJ: Erlbaum; 2004. p. 61–82.
54. Kenneth D. Primero Dios: alcoholics anonymous and the Hispanic community. Selinsgrove: Susquehanna University Press; 1994.
55. Boyd-Franklin N, Garcia-Preto. Family therapy: the cases of African American and Hispanic women. In: Comas-Diaz L, Greene B, editors. Women of color: integrating ethnic and gender identities in psychotherapy. New York: Guilford Press; 1995. p. 239–64.
56. Falicov C. Mexican families. In: McGoldrick M, Pearce JK, Giordano J, editors. Ethnicitv and family therapy. New York: Gardner; 1982. p. 134–63.
57. Mejia D. The development of Mexican-American children. In: Powell GJ, Yamamoto J, Romero A, Morales A, editors. The psychosocial development of minority group children. New York: Brunner Mazel; 1983. p. 77–114.
58. Sue DW, Sue D. Counseling the culturally different: theory and practice. New York: John Wiley & Sons; 1990.
59. Gil RM, Vazquez CN. The Maria paradox. New York, NY: Perigee Books; 1996.
60. Canino, G.J., Burnam A., Caetano R., Helzer J.E., editors Alcoholism in North America, Europe, and Asia. In: The prevalence of alcohol abuse and dependence in two Hispanic communities. New York: Oxford University Press; 1992. p. 131–55.
61. Canino GJ. Alcohol use and misuse among Hispanic women: selected factors, processes, and studies. Int J Addict. 1994;29:1083–100.
62. Welte JW, Barnes GM. Alcohol and other drug use among Hispanics in New York state. Alcohol Clin Exp Res. 1995;19(4):1061–6.
63. Pew Research Center. The shifting religious identity of Latinos in the United States [Internet]. Washington, DC: Pew Research Center; 2014 May 7 [cited 2020 Oct 8]. Available from: https://www.pewforum.org/2014/05/07/theshifting-religious-identity-of-latinos-in-the-united-states/.
64. Arredondo R, Weddige RL, Justice CL, Fitz J. Alcoholism in Mexican-Americans: intervention and treatment. Hosp Community Psychiatry. 1987;38(2):180–3.
65. Rebach H. Alcohol and drug use among American minorities. In: Trimble JE, Bolek CS, Niemcryk SJ, editors. Ethnic and multicultural drug abuse: perspectives on current research. New York: Harrington Park Press; 1992. p. 23–57.

Dementia and Cognitive Disorders in Geriatric Hispanic/Latinos

10

Bernardo Ng and Nancy C. Colimon-Ardila

Introduction

Although the terms "Hispanic" and "Latino" are often used interchangeably in the medical literature, they are not synonyms. The term "Hispanic" refers to people born in a country that was "conquered" by Spaniards and for whom Spanish is their primary language. This applies to most countries in Latin America, except Brazil, the Guianas, Belize, Haiti, and Trinidad and Tobago. The term "Latino" on the other hand is a broader term that refers to people born in a country whose language has evolved from Latin (Romance languages), such as Italy, Spain, France, and all countries in Latin America. Historically, different ethnicities have contributed to the contemporary Hispanic/Latino populations in the USA, making this population very heterogeneous from a biological point of view. Genome-wide analysis (GWA) within this population found that Dominicans and Puerto Ricans showed the highest levels of African ancestry, whereas Colombians have a wider distribution of African versus European GWA enrichment. In contrast, Mexico and Peru were two countries with high densities of native Americans during the pre-Colombian times, as it would be expected individuals from these countries have the highest degree of Native American ancestry [1]. This chapter will use the terms Latino and Hispanic interchangeably.

B. Ng (✉)
Asociación Psiquiátrica Mexicana, Mexico City, Mexico

American Society of Hispanic Psychiatry, Imperial, California, USA
e-mail: bng@sunvalleyb.com

N. C. Colimon-Ardila
Director Operations and Innovation, Centro Geriátrico Nuevo Atardecer, Mexicali, Baja California, Mexico

Asociación Colombiana de Psiquiatría Biológica, Mexicali, Baja California, Mexico

R. Castilla-Puentes, T. Falcone (eds.), *Mental Health for Hispanic Communities*,
https://doi.org/10.1007/978-3-031-13195-0_10

The office of the United States Census published in 2003 that Latinos had become the largest minority in the USA. This fact had been projected for decades, a period that was apparently insufficient to prepare for the public and social needs secondary to the expansion of this minority group. Regarding matters of health, and particularly mental health, issues such as stigma, immigration and aging related stress, violence and alcoholism, and concerns about psychotropic medications remain unresolved up to date [2]. These factors and others including obesity, diabetes, and level of disability are likely to shorten the so-called Hispanic paradox in the next decades. Even though, Latinos do live longer, the excess includes at least 6 years with disability (i.e., sensory, physical, mental, self-care, ability to leave the house), in contrast to 4 years or less, living with disability for non-Hispanic whites [3]. Furthermore, in clinic environments with timely screenings for cognitive decline, Hispanics get diagnosed in average four years earlier in age than non-Hispanic whites [4].

This chapter will focus on factors that have placed Latinos in the USA at greater risk of cognitive disorders, delayed identification, and consequently untimely beginning of treatment; when compared to older adults of other ethnicities. These will be presented in the section of risk factors, especially because it is projected that by 2050, the elderly Hispanic population will represent 20% of the 65-year and older segment of the population in the United States. Furthermore, the reported prevalence of major neurocognitive disorder for the 65–74, 75–84, and $\geq$ 85 age groups is 7.5%, 27.9%, and 62.9% compared with the non-Hispanic Whites of 2.9%, 10.9%, and 30.2%, respectively [5]. Furthermore, Hispanics spend more time living with cognitive impairment when compared with other ethnic groups. According to Garcia et al., in a sample of adults 50 years of age and older, Hispanics spent a larger segment of their late life with cognitive impairment when compared with non-Hispanic Whites and Blacks. In the case of women, the number of years spent living with dementia for non-Hispanic Whites, Blacks, US-born Hispanics, and foreign-born Hispanics was 1.6, 3.9, 4.7, and 6.0 years, respectively; whereas in the case of men, the years with dementia were 1.1, 3.1, 3.0, and 3.2, respectively [6].

Identification of at-risk Hispanic elderly population represents a challenge in both clinical and research settings due to barriers such as language, low educational attainment, low income, and a negative attitude towards participation in research [5]. Therefore, this chapter will close with evidence supported recommendations to address the identified gaps and risk factors, looking at improving outcomes of the elderly Latino population.

Risk Factors

Ethnicity and Race

The terms "race" and "ethnicity" merit some explanation. "Race" primarily alludes to those physical features that are usually determined by genetics. On the other hand, Hispanic and Latino, and the most recently used Latinx, are ethnicity-based terms, since "ethnicity" refers to features that provide a particular distinctiveness to those belonging to a given group of individuals, expressed in language, values, family traditions, religion, art, music, literature, use of names, public life, and culture [1]. Furthermore, any and/or all the features that define an ethnic group, impact the way symptoms and diseases are understood and interpreted by the individuals belonging to such ethnic group, and by the health professionals providing care to them [7–9].

Ethnoracial Studies

Ethnoracial studies of brain structure change and cognitive change have contributed greatly in the understanding of the most likely pathophysiology of cognitive disorders in Latinos. Gavett et al. [10] compared African Americans ($n = 121$), Hispanics ($n = 111$), and non-Hispanic Whites ($n = 212$); using as main outcome measure, MRI hippocampal gray matter (GM) volume and white matter hyperintensity (WMH), in a multivariate model. The cognitive outcomes in this study were composite measures of episodic memory, semantic memory, executive function, and spatial ability. Even though all groups showed a mixture of GM volume loss and WMH, as measured by MRI, there were varied degrees that showed a clear tendency in each group. African Americans had a predominance of global GM volume loss, non-Hispanic Whites had a mix of GM and WMH volume loss, whereas Hispanics had a predominance of WMH ($p < 0.05$).

Although not entirely specific, gray matter atrophy, specifically temporal lobe atrophy is commonly associated with Alzheimer's disease pathology, whereas WMH is a marker of microvascular disease associated with non-AD processes such as type 2 diabetes and cerebrovascular disease. This pivotal finding gives a singular perspective to those involved in the treatment of cognitive disorders in Latinos, especially in the opportunities to intervene preventively. Given its importance, both diabetes and hypertension in Latinos are further addressed in this chapter.

Migration Status

Latinos continue to comprise the largest ethnic group of immigrants to the USA [11]. The phenomenon of immigration has complex effects on cognitive function. Both the educational attainment, and the utilization and type of health systems in the country of origin, in the pre-migration phase; followed by the actual migration process where a number of "selection factors" come in play; and eventually the process of acculturation once the migration process has been completed, must all be considered. Moon et al. studied the level of cognitive function of 7609 older adults, and found that immigrant status moderated the relationship between ethnic and dementia, establishing that US-born Hispanics had a better level cognition than Hispanic immigrants [12].

Both pre- and post-migration-related stress have an impact in the health and mental health of Latinos, either documented or undocumented. In the case of undocumented immigrants that become older adults in this country, they continue to live in the face of anxiety about possible apprehension and deportation. They also carry the burden of having left close family members behind, whom in many cases do not get to see for decades [3].

The phenomenon of migration has also been studied in Latinos with cognitive impairment. A study by Garcia et al., with 20 years of data from the Hispanic Established Populations for the Epidemiologic Study of the Elderly or Hispanic-EPESE, including over 13,000 subjects generated important findings. The total sample was 44% immigrants and 56% US-born, with 58% being women. The main outcome measure was the mini mental status examination, that could be administered either in English or Spanish. This group found that Mexican American immigrants regardless of gender live more years with cognitive impairment after age 65, compared to US-born Mexican Americans, discrepant with the Hispanic paradox. Potential contributors are the lower levels of education of Latino immigrants. They also found that immigrant women were at a greater disadvantage than immigrant men, suggesting that men may be more health selected for migration. Another finding was that US-born Mexican Americans had a later onset of cognitive decline, but once installed the decline would eventually draw near the deficit of immigrants. The authors suggest that the higher educational level and better access to preventive care of the US-born worked as a protective factor [13].

Another finding by the same group, reporting in regard to age of migration, was that those that arrive in mid-life have a healthier cognitive function after age 65, compared to early- and late-life immigrants. The authors highlight the fact that immigrants are a very heterogenous group, and that numerous factors including economic and educational level, gender, and age of migration, impact the level of cognitive decline among Latinos [14].

Education Level

It is well established that Latinos hold a lower educational level in the USA compared to other ethnic groups. As per the US Census, in 2019, 40.1% of non-Hispanic whites age 25 and older had a bachelor's degree or higher, up from 33.2% in 2010. During the same periods, the percentage of African Americans age 25 and older with a bachelor's degree or higher scaled from 19.8% to 26.1%; Asians from 52.4% to 58.1%; and Hispanics from 13.9% to 18.8% [15].

Mungas et al. analyzed data from the SALSA study, where they studied the cognitive trajectories based on level of education, with up to seven assessment waves in a total of nine years. They were able to establish that a higher educational level did have a protective factor on the baseline assessment. By the same token the preferred language, or the status of monolingualism vs. bilingualism did not have an effect [16]. Interestingly enough, a study by Wilson et al. compared the trajectory of cognitive decline for an average of 6 years, in annual assessment of a group of Latinos ($n = 104$) compared with a group of non-Latinos ($n = 104$), matched in educational level, showed no differences in the rate of decline. This study suggests that educational attainment could obliterate the effect of variables such as migration and/or ethnicity itself [17].

Health Literacy

A genetic study, to explore cultural beliefs about Alzheimer disease (AD), included 37 Mexican American family members of patients diagnosed with cognitive impairment, reported findings of interest. First, that the majority believed that AD was a normal part of aging, or it was a mental illness caused by bad habits. Second, very few understood being at risk of inheriting the disease, but once informed, most expressed a strong desire to be tested. The researchers recommended comprehensive counseling about AD and genetic testing. Albeit small, this study points two important features, the commonly known low literacy in cognitive disorders, and more important, the positive response to adequate and comprehensive literacy efforts [18].

On the other hand, interethnic comparison studies have shown that older Latinos are less likely to use computers or other digital technologies, which represents yet another disadvantage when trying to improve educational attainment and health literacy in this ethnic group. If the intention is to take advantage of new information technologies, training in the use of such technologies should be part of the plan [19].

Socioeconomic Status

Socioeconomic status (SES) also plays a role in access to preventive health measures, and impact cognitive decline at later life. Data from the SALSA study that included 1789 community dwelling Latinos, evaluated SES defined as a variable created by level of education, income, and occupation. Cognitive function was measured by the Mini Mental Status Examination and the Spanish English Verbal Learning Test. Their results showed that SES trajectories in the subjects' life span, had an impact, both on global cognitive function and short-term verbal memory. Those subjects with higher SES in childhood and adulthood experienced less global cognitive and verbal memory decline than those with lower SES ($p < 0.0001$). Furthermore, those with an SES trajectory from low to high was also associated with less decline ($p < 0.0001$). This study is significant for the composition of the SES variable, where not only income but occupation and education status were evaluated, which attempts to illustrate how these social determinants hardly operate independent from each other. The results of these studies strongly suggest that a favorable trajectory in SES promotes a benefit to health outcomes decades later, in this case cognitive function in older life. Furthermore, even if SES is low early in life but the trajectory moves to a higher SES, also promotes a benefit in later life [20].

Non-psychiatric Comorbidities

Cardiovascular and metabolic comorbidities are important in the study of cognition, especially hypertension and diabetes. The literature has discussed several potential mechanisms through which heart rate variability (HRV) may influence brain structure and function. The baroreflex mechanism regulates blood flow and maintains proper perfusion to vital organs, including the brain, through modulation of the heart rate and contractibility. In other words, cardiac autonomic function and the sympathetic and parasympathetic activity interact to maintain blood pressure within a normal range. In fact, HRV and blood pressure variability (BPV) have been shown to be inversely associated. Furthermore, fluctuations in blood pressure (i.e., BPV) are associated with cognitive impairment and with structural brain changes related to hypertension, such as cerebral white matter lesions, and lacunar infarctions. Additionally, HRV may influence cognitive function through cardiac autonomic neuropathy and its associations with diabetes, impaired plasma glucose, and insulin sensitivity. A study of 869 elderly Latinos by the University of California, San Francisco, demonstrated that those with low HRV scored 4 points lower than those with normal HRV in a modified version of the Mini Mental Status Exam ($p < 0.01$), supporting the hypothesis of diabetic-related cardiac autonomic neuropathy [21].

By the same token, it is well known that Latinos in the USA have a disproportionately higher prevalence of type 2 diabetes, compared to other ethnic groups. According to a recent study by Cheng et al., the weighted age-and-sex adjusted prevalence for Hispanic adults in the USA is of 22.1%, compared to 12.1% for non-Hispanic white, 20.4% for non-Hispanic Black, and 19.1% for non-Hispanic Asian adults ($p < 0.001$). Within the different groups of Hispanics, the prevalence was 24.6% for Mexicans, 21.7% for Puerto Rican, 20.5% for Cuban/Dominican, 19.3% for Central American, and 12.3% for South American subgroups ($p < 0.001$) [22].

The group of Strizich et al. was able to collect data supporting the hypothetical notion that the relationship between type 2 diabetes and cognitive impairment may be bidirectional, based on results of neuropsychological testing. They administered the DSST and the Word Fluency and Learning delayed recall components of the Spanish English Verbal Learning Test, to 1794 Hispanic diabetics ages 45 to 76. In this study they used HbA1c ≥ 7% to identify cases of uncontrolled diabetes. Besides the neuropsychological tests, the group administered questionnaires covering sociodemographic, medical, environmental, and lifestyle components. After adjustment, lower DSST scores showed an association with poorly controlled diabetes ($p = 0.03$); interestingly enough this result was accentuated in cases of poor family support. Among those with low family support, persons in the lower quartile of global cognitive function were more than twice as likely to have uncontrolled diabetes, than those in the highest performing quartile (OR = 2.31; 95% CI:1.17–4.55); there was no comparable effect among those with high family support. The latter suggested that high family support appeared to attenuate the effect [23].

On the other hand, Bangen et al. examined the progression of cognitive decline in a multiethnic sample of 1493 older adults (≥65 years of age), that included 346 diabetic and 151 non-diabetic Latinos. The subjects were assessed at baseline with a neuropsychological battery that included measurements of memory, language, processing speed, executive functioning, and visuospatial abilities. The assessment was repeated every 18 months for an average of 6 years. Although the rate of progression of cognitive decline was similar between the groups, the cognitive performance of the diabetic sample performed significantly more poorly ($p < 0.05$) from baseline and all follow-up measurements, relative to those without diabetes. The researchers suggest that diabetes impacts cognition since the onset of the metabolic dysfunction, and by age 65 the cognitive performance measured by this neuropsychological battery exhibits compromised function, persistently poorer than the cognitive performance of non-diabetics of the same age [24].

Cardiovascular disease and cardiovascular risk factors including dyslipidemia (i.e., total cholesterol ≥240 mg/dL, LDL cholesterol ≥160 mg/dL, or HDL ≤40 mg/dL, or receiving cholesterol-lowering medication), hypertension (systolic blood pressure ≥ 140 mm Hg, diastolic blood pressure ≥ 90 mm Hg, or receiving antihypertensive medication), and obesity (i.e., ≥BMI of 30.0) were explored differentially by Daviglus et al. among 15,079 Latinos of diverse backgrounds. Dyslipidemia was present in 52% of men, especially represented by Dominican and Puerto Rican; and 37% of women, particularly Puerto Rican. Hypertension was present in 25% of men, highest among Dominican, and 24% in women with Puerto Rican being more affected. Obesity was present in 37% of men and 43% of women, in both cases mostly Puerto Rican. About 15% of men and women were using antihypertensive medication. The prevalence of coronary heart disease was 4% for men and 2% for women, whereas stroke was 2% for men and 1% for women [25].

Although the evidence above comes from different studies, it is presented to illustrate the challenge for those involved in treating cognitive disorders in older Latinos. Clinicians and researchers must keep in mind that a quarter of adult Latinos in the USA has diabetes, a half dyslipidemia, and about a fifth deal with hypertension, coronary heart disease, and stroke. More important yet, all these brain-related factors that increase the risk of developing cognitive disorders are preventable.

Depressive Symptomatology

Depressive symptoms will be addressed from two viewpoints, on the one hand depressive symptoms, stress, and unhealthy behaviors; and on the other hand, as manifestations of a degenerative disorder.

On the first aspect, Rodriquez et al. studied a multiethnic sample ($N = 6479$) of adults between ages 50 and 59, to see the development of depressive symptoms, measured by the CES-D, as a result of the interaction of chronic stress and unhealthy behaviors, in a period of two years. This study defined as chronic stress factors, conditions such as serious physical attacks, accidental injury, life-threatening illnesses; and as unhealthy behaviors, obesity as a proxy for poor diet and physical inactivity, use of alcohol, and/or tobacco. Latinos (30.7%) compared to non-Latino Whites (16.8%) and African Americans (21.0%), had higher rates of depressive symptoms ($p < 0.001$), but not as many stress factors. As to the unhealthy behaviors, Latinos main risk factor for future development of depressive symptoms was mild obesity (OR 2.5), which also represents a risk factor for cognitive impairment. In this study, educational attainment was a secondary outcome, and in all cases a higher level was a protective factor [26].

Data from the ¡Caminemos! study analyzed by Hernandez et al. showed that the presence of depressive symptoms was significantly and negatively associated with cognitive function. This study included 572 participants where 77.1% were women and had a mean age of 77.3 (SD = 6.76) years. However, the association was lost at

1-year and 2-year follow-up assessments. At baseline, participants that screened positive to depression were 1.58 times more likely to experience cognitive impairment ($p = 0.04$). Subjects identified for this study were sedentary at baseline, and recruited to participate in an exercise program (see below), while measuring both depressive and cognitive symptoms until the completion of the study. The study results suggest that physical exercise can improve both symptoms, suggesting a bidirectional relationship of both clinical conditions [27].

Finally, Guerrero-Berroa et al. [5] studied the weight of different depressive symptoms as predictors for cognitive disorders, in a sample of 4334 subjects, comparing Hispanics vs. non-Hispanic Whites. Their findings support the idea that apathy, lack of motivation, anhedonia, and mental slowness measured in depression scales are associated with brain dysfunction and major cognitive disorders, and not only a mood disorder.

In summary, depressive symptoms in Latino older adults interact with cognitive symptoms in complex ways. Symptoms of major depressive disorder, especially apathy and mental slowness may actually be an early manifestation of a cognitive disorder.

Barriers to Care

One of the main barriers to care is the deficiency of clinicians with sufficient cultural competence to administer tests on Latinos on the one hand, and the availability of instruments in Spanish on the other hand [28].

In regard to the available instruments in Spanish, batteries for cognitive testing have been translated and adapted to "neutral" Spanish, with reliable psychometric properties in episodic memory, working memory, semantic memory, perceptual speed, and visuospatial ability, for over a decade. This adaptation by Krueger et al. was done with community dwelling elderly adults between 60 and 75 years of age, showing that with a trained clinician the test results are reliable. However, the group also noted as a limitation that their results could not be generalized to elderly Latino individuals in different living conditions, such as nursing or retirement homes [29].

Recommendations

Literacy Campaigns

Health literacy efforts with good results have already been tested, these include the use of infographics, "fotonovela," and pictograms. Sixteen caregivers, adult children, and/or spouses were included in a literacy program for caregivers with good results. The study demonstrated that the use of infographics that can be tailored to the caregivers needs and situation can enhance the understanding of the disease and improve self-care measures. A fotonovela (FN) named "Together We Can! Facing memory loss as a family" gave good results when tested during a study of 110 caregivers. They were exposed to either FN or information-as-usual-material, to get educated on self-care and identification of dementia symptoms. The group that received the FN intervention had a larger improvement in depressive symptoms, after a six-month follow-up assessment [30–32].

Training of Clinicians

Teletesting either by phone or video meeting, arises as an obvious solution to overcome the shortage of qualified clinicians. Testing by phone is not new and has been examined in Latinos with the Dementia Questionnaire (DQ) and the Telephone Interview for Cognitive Status (TICS) to identify Major Neurocognitive Disorder (MaND) and Minor Neurocognitive Disorder (MiND), respectively [33]. In this study, the assessments by phone correctly identified 66.5% of participants with normal cognition, 55.4% of those with MiND, and 92.7% of those with MaND.

Proper training is also important, since a study of older Hispanics complaining of subjective cognitive decline (SCD) were more likely to be depressed when objectively screened for memory impairment. According to Mainly et al. (2011), who studied 145 Hispanic subjects in California, with the Mini Mental Status Examination, Wechsler Memory Scale-Revised Logical Memory Test, the Consortium to Establish a Registry for Alzheimer's Disease (CERAD) Word List Memory Task, the Clock Drawing Test, the Trail Making Test (parts A and B), and a Category Fluency test, and the Geriatric Depression Scale (GDS). A multiple regression analysis showed a significant relationship of SCD with GDS scores ($p = 0.001$), and not with Global Cognitive Composite scores ($p = 0.066$) or Memory Composite scores ($p = 0.936$) [33].

These data underscore the importance of appropriate training, in order to have the necesary compentence to work through the clinical challenge of differentiating major depressive from major neurocognitive disorders.

Optimize Medication Adherence

According to Mohan et al. (2014), the use of "PictureRx illustrated" medication list depicting each medication's indication and dosing instructions, improved overall understanding of their medications ($p < 0.001$), including greater ability to report the drug indication ($p < 0.01$), strength ($p < 0.05$), dosing ($p < 0.01$), and frequency of administration ($p < 0.001$), in a group of 200 Latinos [34]. From the point of view of cognitive impairment, the goal would be to optimize adherence to dementia medication, as well as comorbidities like diabetes and hypertension [35, 36].

Participation in Clinical Trials

Also important, and of great long-term value, is promoting the increased participation of Latinos in clinical trials that are on the search of new pharmacological options for the treatment of neurocognitive disorders [37]. This should include support to clinical research sites located in communities of predominantly Latino populations, especially those located in remote areas [38].

Physical Exercise and Dietary Adjustments

Physical activity has demonstrated benefit in muscle strength, balance, aerobic power, endurance, and mobility. It has also been proposed as protective against negative effects of aging on the brain. The study ¡Caminemos! including Latino individuals 60 years or older ($N = 571$) by the group of Piedra et al. demonstrated that regular physical exercise can prevent deterioration and even improve cognitive function, for as long as two years. Improvement was statistically significant after the first year ($p = 0.001$) and the second year ($p = 0.013$) when compared to baseline. In their study, the exercise program consisted of a series of four, weekly 1-hour sessions of exercise class targeting muscle strength, endurance, balance, and flexibility for one month, which were based on a modified version of the Lifetime Fitness Program. They were followed by monthly reinforcement sessions for 11 additional months, and bi-monthly sessions for 12 more months, which completed the 24 months of the study. These exercise sessions were implemented after an intervention session on attribution-retraining (treatment group) or a health education session (control group). The benefit was not significantly different between the groups ($p = 0.582$), suggesting that exercise offered an advantage regardless of the intervention. It is important to mention that people who were already physically active at least three times a week were not included in this study [39].

This study illustrates at two very important issues to be considered as recommendations, first that physical exercise benefits cognitive function in elderly Latinos, and second that Latinos do engage in physical exercise interventions.

A study of 2702 Mexican Americans from Texas demonstrated that those participants who met recommendations of five or more servings of fruit and vegetable per day had 58% decreased odds of mental disorders symptoms (OR = 0.42; 85% CI: 0.18–0.97), as measured with the Mini Mental Status Exam, the Centers for Epidemiologic Studies Scale in English and Spanish, and the Zung's self-Rating Anxiety Scale [40]. This is of great importance, since historically the origin of Mexican food and that from other Latin countries consists of herbs and vegetables that are not generally recognized in current Latino meals in the USA [41].

Regular physical exercise and dietary adjustments can clearly help prevent and improve cognitive impairment both in the short and long term. At the same time the same measures can improve outcomes in other risk factors already reviewed, such as obesity, hypertension, type 2 diabetes, that directly and indirectly can prevent cognitive impairment.

Community-Based Adult Services

Community-Based Adult Services (CBAS) or as previously known Adult Day Healthcare Centers (ADHC) are intended to offer services that are expected to close the gap of care needed by the elderly population that requires a level higher than outpatient services, but do not meet criteria for services at a nursing skilled facility. They have been available in the USA and other countries for decades, and are especially directed to the population who are identified as cognitively impaired/no dementia (CIND), these are people with measurable impairment but without measurable impact in function. Research has demonstrated that the services provided at these programs can delay the onset of cognitive disorder [6]. Based on the fact that Latinos tend to live more years in CIND status, strategies should target specific needs for this population segment with culturally appropriate programs. These programs have been associated with improvement in cognitive function and depressive symptoms and include professional nursing services; physical, occupational and speech therapies; mental health services; therapeutic activities, such as listening to music, physical activity, dancing, and solving puzzles; social services; personal care; a meal and nutritional counseling; transportation to and from the participant's residence and the CBAS center [19, 42].

Properly trained personnel at these centers can assist in a better understanding of prevention strategies in cognitive, mood, metabolic, and cardiovascular disorders, as well as become the avenue to educate the elderly and their families in genetic aspects of brain degenerative diseases [43–45].

Caring for the Caregivers

It is generally believed that due to the widely identified familism among Latinos, care for the elderly would be guaranteed. However, social forces (i.e., financial, family disruption) prevent some families to provide care to the full extent. There are at least three variations of Latino families identified [46]:

(a) Extensive, defined as the one provided by the entire family both with instrumental and emotional support. Caregivers with extensive support usually have a larger family size, adaptable family members, help outside of the family, and formalized processes for spreading caregiving duties across multiple persons.
(b) Limited, includes instrumental support provided by only one family member, who is at high risk for burnout and distress. In this scenario the dementia-related neuropsychiatric symptoms are frequently seen as obstacles to family unity.
(c) Absent, is the complete lack of support, and have the worst outcomes, requiring institutional services very early.

The study cited above by Garcia et al. also found that cognitive impairment especially when complicated by neuropsychiatric symptoms, is an important predictor for family caregiver psychological distress [13].

According to Rote et al., culturally competent interventions should take into consideration diversity in Latino dementia care by providing psychoeducation on problem solving and communication skills to multiple family members, particularly with respect to the natural history of cognitive disorders and neuropsychiatric symptoms, and by assisting caregivers in managing family tensions and disruption, with the goal to mobilize support when needed [46]. Regardless of the type of family caring for a Latino elderly patient, proper care cannot be provided without including the family members in the treatment plan. This is a premise while applicable to families of any ethnic background, is extremely important in Latino families.

COVID-19 Pandemic

During the writing of this chapter, we were in the middle of the COVID-19 pandemic, so it was considered appropriate to include a comment. Given the data reviewed so far, it is evident that the Hispanic/Latino geriatric population is at an unfortunate disadvantage. Among the independent risk factors associated with increased COVID-19 mortality are history of diabetes, hypertension, and older age [47]. Among the independent factors for advanced medical care (i.e., mechanical ventilation) is the presence of obesity (BMI > 35 kg/m^2) [48]. As mentioned through this chapter, all these factors are overrepresented in the Hispanic elderly population.

On the other hand, neuroinflammation related to metabolic and cardiovascular factors, especially related to type 2 diabetes, is a prominent feature in brain degenerative disease in Latinos. Proposed immunological activation and excessive inflammation related to COVID-19 may fast-track the progression of pre-existing brain inflammatory neurodegeneration during this pandemic, making elderly individuals more susceptible to adverse outcomes when contracting SARS-CoV-2 infection. It is therefore reasonable to argue that the elderly Hispanic in the USA, when compared to other ethnic groups, are at an increased risk for neurocognitive disorders, as well as severe outcomes after SARS-CoV-2 infection [49].

It is plausible to consider then that the convergent genetic, socioeconomic, comorbid (type 2 diabetes) factors recognized as linked to COVID-19 infection severity may create an exceptionally high-risk profile for Latinos for poor outcomes during this pandemic [50].

It is worth mentioning that Adult Day Care Centers across the country have been instrumental in providing aid to the Latino elderly and their families through telehealth services. This has helped with treatment adherence, prescriptions refills, and continued contact with their providers during this public health disaster, repeating what has been achieved in previous emergencies, such as the 2010 earthquake in California [51, 52].

Conclusions

This chapter has presented information unique to the Latino elderly with cognitive impairment and cognitive disorders. It is our opinion that there are many open and pending issues to address, especially socioeconomic status and educational attainment as social determinants of these disorders. Even though change in these areas may take a long time, in the short term the use of proven methods to improve the level of health literacy, medication adherence, in a friendly and non-threatening way shall permit the elderly Latino and his/her families, feel confident and at ease to seek advice and care in a timely manner. One of the most important features to address is generating mechanisms to enhance the understanding of the differences between normal aging and pathological cognitive decline.

Another outstanding feature in this chapter is the great opportunity of preventive programs to address the physiological, dietary, and metabolic issues, that in the long run have an impact on cognitive function. It is foreseeable that strategic collaboration among the neuroscientific, metabolic, and cardiovascular fields can optimize resources and have an impact in longevity, quality of life, and eventually reduce the number of years with cognitive impairment, at a lower cost.

It is our expectation that the material here presented is useful to clinicians and non-clinicians in the area of neurosciences, either Latinos or not, and it will hopefully trigger the curiosity to learn more about neurocognitive disorders in the Hispanic/Latino elderly population.

References

1. Aguayo-Mazzucato C, Diaque P, Hernandez S, Rosas S, Kostic A, Caballero AE. Understanding the growing epidemic of type 2 diabetes in the Hispanic population living in the United States. Diabetes Metab Res Rev. 2019 Feb;35(2):e3097. https://doi.org/10.1002/dmrr.3097.
2. Watson MR, Kaltman S, Townsend TG, Goode T, Campoli M. A collaborative mental Health Research agenda in a community of poor and underserved Latinos. J Health Care Poor Underserved. 2013 May;24(2):671–87. https://doi.org/10.1353/hpu.2013.0079.
3. Goldman N. Will the Latino Mortality advantage endure? Res Aging. 2016. Apr;38(3):263–82. https://doi.org/10.1177/0164027515620242.
4. Fitten LJ, Ortiz F, Fairbanks L, Bartzokis G, Lu P, Klein E, Coppola G. Ringman JYounger age of dementia diagnosis in a Hispanic population in southern California. Int J Geriatr Psychiatry. 2014 Jun;29(6):586–93. https://doi.org/10.1002/gps.4040.
5. Guerrero-Berroa E, Kluger A, Schmeidler J, Sailor K, Lizardi H, Golomb J, Ferris S, Reisberg B. Neuropsychological and neuropsychiatric prediction of global cognitive status among older Spanish-speaking Hispanics and English-speaking whites. J Geriatr Psychiatry Neurol. 2014 Dec;27(4):266–75. https://doi.org/10.1177/0891988714532020.
6. Garcia MA, Downer B, Chiu CT, Saenz JL, Rote S, Wong R. Racial/ethnic and nativity differences in cognitive life expectancies among older adults in the United States. Gerontologist. 2019;59(2):281–9. https://doi.org/10.1093/geront/gnx142.
7. Rosenbloom JM, Burns SM, Kim E, August DA, Ortiz VE, Houle TT. Race/ethnicity and sex and opioid Administration in the Emergency Room. Anesth Analg. 2019;128(5):1005–12.
8. Braksmajer A, Fedor TM, Chen SR, Corales R, Holt S, Valenti W, McMahon JM. Willingness to take PrEP for HIV prevention: the combined effects of race/ethnicity and provider trust. AIDS Educ Prev. 2018 Feb;30(1):1–12.
9. Ng B, Dimsdale JE, Shragg GP, Deutsch R. Ethnic differences in analgesic consumption for post-operative pain. Psychosom Med. 1996;58:125–9.
10. Gavett BE, Fletcher E, Harvey D, Farias ST, Olichney J, Beckett L, DeCarli C, Mungas D. Ethnoracial differences in brain structure change and cognitive change. Neuropsychology. 2018 Jul;32(5):529–40. https://doi.org/10.1037/neu0000452.
11. Largest immigrant group. https://www.migrationpolicy.org/programs/data-hub/charts/largest-immigrant-groups-over-time (consulted 06/12/2020).
12. Moon H, Badana ANS, Hwang SY, Sears JS, Haley WE. Dementia prevalence in older adults: variation by race/ethnicity and immigrant status. Am J Geriatr Psychiatry. 2019;27(3):241–50. https://doi.org/10.1016/j.jagp.2018.11.003. Epub 2018 Nov 28
13. Garcia MA, Saenz JL, Downer B, Chiu CT, Rote S, Wong R. Age of migration differentials in life expectancy with cognitive impairment: 20-year findings from the Hispanic-EPESE. Gerontologist. 2018 Sep 14;58(5):894–903. https://doi.org/10.1093/geront/gnx062.
14. Garcia MA, Ortiz K, Arévalo SP, Diminich ED, Briceño E, Vega IE, Tarraf W. Age of migration and cognitive function among older Latinos in the United States. J Alzheimers Dis. 2020 Jul 8; https://doi.org/10.3233/JAD-191296.

15. Educational attainment US census. https://www.census.gov/newsroom/press-releases/2020/educational-attainment.html#:~:text=In%202019%2C%2040.1%25%20of%20non,from%2013.9%25%20to%2018.8%25. (Consulted 08/09/2020).
16. Mungas D, Early DR, Glymour MM, Zeki Al Hazzouri A, Haan MN. Education, bilingualism, and cognitive trajectories: sacramento area Latino aging study (SALSA). Neuropsychology. 2018 Jan;32(1):77–88. https://doi.org/10.1037/neu0000356.
17. Wilson RS, Capuano AW, Marquez DX, Amofa P, Barnes LL, Bennett DA. Change in Cognitive Abilities in Older Latinos. J Int Neuropsychol Soc. 2016 Jan;22(1):58–65. https://doi.org/10.1017/S1355617715001058.
18. Withers M, Sayegh P, Rodriguez-Agudelo Y, Ernstrom K, Raman R, Montoya L, Zuno-Reyes A, Mosieri C, Matute E, Ringman JM. A mixed-methods study of cultural beliefs about dementia and genetic testing among Mexicans and Mexican-Americans at-risk for autosomal dominant Alzheimer's disease. J Genet Couns. 2019 Oct;28(5):921–32. https://doi.org/10.1002/jgc4.1133.
19. Herrera AP, Meeks TW, Dawes SE, Hernandez DM, Thompson WK, Sommerfeld DH, Allison MA, Jeste DV. Emotional and cognitive health correlates of leisure activities in older Latino and Caucasian women. Psychol Health Med. 2011;16(6):661–74. https://doi.org/10.1080/13548506.2011.555773.
20. Haan MN, Zeki Al-Hazzouri A, Aiello AE. Life-span socioeconomic trajectory, nativity, and cognitive aging in Mexican Americans: the Sacramento area Latino study on aging. J Gerontol B Psychol Sci Soc Sci. 2011 Jul;66(Suppl 1):i102–10. https://doi.org/10.1093/geronb/gbq071.
21. Al Hazzouri AZ, Haan MN, Deng Y, Neuhaus J, Yaffe K. Reduced heart rate variability is associated with worse cognitive performance in elderly Mexican Americans. Hypertension. 2014 Jan;63(1):181–7. https://doi.org/10.1161/HYPERTENSIONAHA.113.01888.
22. Cheng YJ, Kanaya AM, Araneta MRG, Saydah SH, Kahn HS, Gregg EW, Fujimoto WY, Imperatore G. Prevalence of diabetes by race and ethnicity in the United States, 2011-2016. JAMA. 2019 Dec 24;322(24):2389–98. https://doi.org/10.1001/jama.2019.19365.
23. Strizich G, Kaplan RC, González HM, Daviglus ML, Giachello AL, Teng Y, Lipton RB, Grober E. Glycemic control, cognitive function, and family support among middle-aged and older Hispanics with diabetes: the Hispanic community health study/study of Latinos. Diabetes Res Clin Pract. 2016 Jul;117:64–73. https://doi.org/10.1016/j.diabres.2016.04.052.
24. Bangen KJ, Gu Y, Gross AL, Schneider BC, Skinner JC, Benitez A, Sachs BC, Shih R, Sisco S, Schupf N, Mayeux R, Manly JJ, Luchsinger JA. Relationship between type 2 diabetes mellitus and cognitive change in a multiethnic elderly cohort. J Am Geriatr Soc. 2015;63(6):1075–83. https://doi.org/10.1111/jgs.13441. Epub 2015 Jun 11
25. Daviglus ML, Talavera GA, Avilés-Santa ML, Allison M, Cai J, Criqui MH, Gellman M, Giachello AL, Gouskova N, Kaplan RC, LaVange L, Penedo F, Perreira K, Pirzada A, Schneiderman N, Wassertheil-Smoller S, Sorlie PD, Stamler J. Prevalence of major cardiovascular risk factors and cardiovascular diseases among Hispanic/Latino individuals of diverse backgrounds in the United States. JAMA. 2012 Nov. 7;308(17):1775–84. https://doi.org/10.1001/jama.2012.14517.
26. Rodriquez EJ, Gregorich SE, Livaudais-Toman J, Pérez-Stable EJ. Coping with chronic stress by unhealthy behaviors: a re-evaluation among older adults by race/ethnicity. J Aging Health. 2017;29(5):805–25. https://doi.org/10.1177/0898264316645548. Epub 2016 May 12
27. Hernandez R, Cheung E, Liao M, Boughton SW, Tito LG, Sarkisian C. The association between depressive symptoms and cognitive functioning in older Hispanic/Latino adults enrolled in an exercise intervention: results from the "¡Caminemos!". Study J Aging Health. 2018 Jul;30(6):843–62. https://doi.org/10.1177/0898264317696776.
28. Ramirez-Gomez L, Jain FA, D'Orazio LM. Assessment of the Hispanic cognitively impaired elderly patient. Review Neurol Clin. 2017 May;35(2):207–29. https://doi.org/10.1016/j.ncl.2017.01.003.

29. Krueger KR, Wilson RS, Bennett DA, Aggarwall NT. A battery of tests of assessing cognitive function in older Latino persons. Alzheimer Dis Assoc Disord. 2009;23(4):384–8. https://doi.org/10.1097/WAD.0b013e31819e0bfc.
30. Arcia A, Suero-Tejeda N, Bakken S. Development of pictograms for an interactive web application to help hispanic caregivers learn about the functional stages of Dementia.
31. Arcia A, Suero-Tejeda N, Spiegel-Gotsch N, Luchsinger JA, Mittelman M, Bakken S. Helping Hispanic family caregivers of persons with dementia "get the picture" about health status through tailored infographics. Gerontologist. 2019 Sep 17;59(5):e479–89. https://doi.org/10.1093/geront/gnz085.
32. Gallagher-Thompson D, Tzuang M, Hinton L, Alvarez P, Rengifo J, Valverde I, Chen N, Emrani T, Thompson LW. Effectiveness of a fotonovela for reducing depression and stress in Latino dementia family caregivers. Alzheimer Dis Assoc Disord. 2015 Apr-Jun;29(2):146–53. https://doi.org/10.1097/WAD.0000000000000077.
33. Manly JJ, Schupf N, Stern Y, Brickman AM, Tang MX, Mayeux R. Telephone-based identification of mild cognitive impairment and dementia in a multicultural cohort. Arch Neurol. 2011 May;68(5):607–14. https://doi.org/10.1001/archneurol.2011.88.
34. Zlatar ZZ, Muniz MC, Espinoza SG, Gratianne R, Gollan TH, Galasko D, Salmon DP. Subjective cognitive decline, objective cognition, and depression in older Hispanics screened for memory impairment. J Alzheimers Dis. 2018;63(3):949–56. https://doi.org/10.3233/JAD-170865.
35. Mohan A, Riley B, Schmotzer B, Boyington DR, Kripalani S. Improving medication understanding among Latinos through illustrated medication lists. Am J Manag Care. 2014 Dec 1;20(12):e547–55.
36. Chandler J, Sox L, Kellam K, Feder L, Nemeth L, Treiber F. Impact of a culturally tailored mHealth medication regimen self-management program upon blood pressure among hypertensive Hispanic adults. Int J Environ Res Public Health. 2019 Apr 6;16(7):1226. https://doi.org/10.3390/ijerph16071226.
37. Cooper C, Tandy AR, Balamurali TB, Livingston G. A systematic review and meta-analysis of ethnic differences in use of dementia treatment, care, and research. Am J Geriatr Psychiatry. 2010 Mar;18(3):193–203. https://doi.org/10.1097/JGP.0b013e3181bf9caf.
38. Ng B, Camacho A, Dimsdale J. Rural psychiatrists creating value for academic institutions. Frontline Reports Psychiatric Services. 2013;64(11):1177–8.
39. Piedra LM, Andrade FCD, Hernandez R, Boughton SW, Trejo L, Sarkisian CA. The influence of exercise on cognitive function in older Hispanic/Latino adults: results from the "¡Caminemos!". Study Gerontologist. 2017 Nov 10;57(6):1072–83. https://doi.org/10.1093/geront/gnw256.
40. Wu S, Fisher-Hoch SP, Reininger BM, McCormick JB. Association between fruit and vegetable intake and symptoms of mental health conditions in Mexican Americans. Health Psychol. 2018;37(11):1059–66. https://doi.org/10.1037/hea0000646. Epub 2018 Oct 8
41. Mexican Food History. https://world-food-and-wine.com/mexican-food-history#:~:text=The%20history%20of%20Mexican%20food,%2C%20tropic%20fruits%2C%20and%20fish. (consulted 8/30/2020).
42. Community Based Adult Services. https://www.dhcs.ca.gov/services/Pages/Community-BasedAdultServices(CBAS)AdultDayHealthCare(ADHC)Transition.aspx (consulted 09/09/2020).
43. Gutierrez AP, McCurley JL, Roesch SC, Gonzalez P, Castañeda SF, Penedo FJ, Gallo LC. Fatalism and hypertension prevalence, awareness, treatment and control in US Hispanics/Latinos: results from HCHS/SOL sociocultural ancillary study. J Behav Med. 2017 Apr;40(2):271–80. https://doi.org/10.1007/s10865-016-9779-x.
44. Ward JB, Albrecht SS, Robinson WR, Pence BW, Maselko J, Haan MN, Aiello AE. Neighborhood language isolation and depressive symptoms among elderly U.S. Latinos Ann Epidemiol. 2018 Nov;28(11):774–82. https://doi.org/10.1016/j.annepidem.2018.08.009.

45. Mosieri C, Matute E, Ringman JM. A mixed-methods study of cultural beliefs about dementia and genetic testing among Mexicans and Mexican-Americans at-risk for autosomal dominant Alzheimer's disease. J Genet Couns. 2019;28(5):921–32. https://doi.org/10.1002/jgc4.1133. Epub 2019 Jun 17
46. Rote S, Angel J, Hinton L. Characteristics and consequences of family support in Latino dementia care. J Cross Cult Gerontol. 2019;34(4):337–54. https://doi.org/10.1007/s10823-019-09378-4.
47. Yu C, Lei Q, Li W, Wang X, Liu W, Fan X, Li W. Clinical characteristics, associated factors, and predicting COVID-19 mortality risk: a retrospective study in Wuhan. China Am J Prev Med. 2020;59(2):168–75. https://doi.org/10.1016/j.amepre.2020.05.002. Epub 2020 May 27
48. Tamara A, Tahapary DL. Obesity as a predictor for a poor prognosis of COVID-19: a systematic review. Diabetes Metab Syndr. 2020;14(4):655–9. https://doi.org/10.1016/j.dsx.2020.05.020. Epub 2020 May 12
49. Abboud H, Abboud FZ, Kharbouch H, Arkha Y, El Abbadi N, El Ouahabi A. COVID-19 and SARS-Cov-2 infection: pathophysiology and clinical effects on the nervous system. World Neurosurg. 2020 Aug;140:49–53. https://doi.org/10.1016/j.wneu.2020.05.193.
50. Naughton SX, Raval U, Pasinetti GM. Potential novel role of COVID-19 in Alzheimer's disease and preventative mitigation strategies. J Alzheimers Dis. 2020;76(1):21–5. https://doi.org/10.3233/JAD-200537.
51. Adult day treatment Centers. https://www.caads.org/adultdday/adultday.html (consulted 8/22/2020).
52. Ng B, Verdugo B, Vahia I. A role of adult day health care centers in earthquake relief efforts. Psychiatr Serv. 2012;63(12):1264.

Mental Health Issues in Latinx/Hispanic Children and Adolescents

11

Tatiana Falcone and Anjali Dagar

Introduction

The prevalence of mental health disorders in children varies according to the different instruments used to measure them; studies have ranged from 1% to 51% according to the publication chosen [1–3]. There is also the issue of how many of the same instruments are validated and appropriately translated to the language; significant differences have been observed when comparing the prevalence of mental disorders in developed countries vs. developing countries [4]. In Latin American countries, the reported prevalence of psychiatric disorders during childhood has been reported with immense variability depending on the methodological differences ranging from 5% to 22% [5]. For example, in a Chilean study, the prevalence of mental health disorders during childhood was reported 22.5% [4]; in a Puerto Rican study, 19.8%.

When assessing the prevalence of mental health disorders in developing countries, it is important to consider the role of poverty and how the different social determinants of health might impact the prevalence rates [6–8]. It is also important to consider the role that stigma can play in Latin American countries accessing appropriate mental health resources, which will delay early diagnosis and prolong mental suffering because patients will be unable to access proper diagnosis and treatment.

Another vital factor to consider is the impact of trauma in immigrant populations. Immigrant youth compared to US origin youth are more likely to have been exposed to community violence and traumatic grief. Some studies in this population have reported refugee Hispanic youth to have lower rates of substance abuse and oppositional defiant disorders [9].

T. Falcone (✉) · A. Dagar
Department of Psychiatry, Cleveland Clinic Lerner College of Medicine of Case Western Reserve University, Cleveland Clinic, Cleveland, OH, USA
e-mail: Falcont1@ccf.org; dagara@ccf.org

R. Castilla-Puentes, T. Falcone (eds.), *Mental Health for Hispanic Communities*,
https://doi.org/10.1007/978-3-031-13195-0_11

Latinx youth are the fastest-growing population in the United States, representing around 25% of all children in the USA [10–12]. Even though they are group as a homogeneous population, Latinx youth originate from 20 different countries, they represent a very heterogeneous group with multiple diverse needs [13].

Depression in Latinx Youth

Sadly to report, Latinx youth have the highest rates of depression among minority groups. A big percentage of Latinx youth in crucial need of mental health services remained underdiagnosed and undertreated, with very limited access to appropriate mental health services in the culturally and linguistically appropriate community [14].

For those Latinx youth who were foreign-born and recently immigrated to the USA, the immigration process's pressure for them and their family can have a deleterious impact on their mental health [15]. Depending on the immigration circumstances, for example those exposed to society violence, poverty, violence exposure, and displacement can certainly impact the development of depression [16]. In the last 10 years, the USA has seen an unprecedented amount of displaced youth from Mexico, Honduras, Nicaragua, and El Salvador secondary to being displaced by violence and in many of the cases remain in the country as unaccompanied minor since 2014 [17, 18]. In recent years, natural disasters have led to the displacement of many Puerto Ricans to the USA [16].

Depression, one of the most frequently recognized sources of disability, is frequently under recognized and undertreated in youth. The longer an episode of depression last, the most difficult it is to treat [19]. Compared with non-Latinx youth, Latinx have higher levels of sadness and hopelessness [20]. Several studies have reported that Latino youth have higher rates of depression and suicide attempts than non-Latino US youth [21]. The role of culture plays an important factor as observed by the fact that US-born Latino youth have consistently high rates of depression and suicide attempts compared to the non-US-born Latino youth [22]. When looking at the gender issue, Latinas are twice more likely compared to Latino men to have suicidal thought, have a suicidal plan or suicide attempts in the last 12 months [20].

Inequities accessing mental health care are highly prevalent in Latinx youth [23]. Studies have demonstrated that Latinx youth are less likely to receive mental health services compared to non-Hispanic white youth [24]; they are also less likely to receive psychotherapy and to be prescribed antidepressants [25]. Financial barriers are another important factors to consider; lack of insurance coverage is something that disproportionally affects Latinx youth's ability to access competent mental health care [26].

Other important social variables to consider for the development of depression in Latinx youth are discrimination, acculturation, and the stress of belonging to a new culture when they want to continue to be part of their original culture; all these factors might adversely impact the development of depression in Latinx youth [27–29]. Language barriers in the kid or the family are an essential component to be considered when assessing Latinx youth. Some studies have reported that for those youth who become the language broker for their family, this can be perceived by them as a burden and increased their risk of alcohol and marijuana use [30]. Also, it has been linked with an increased risk of depression [31]. When there is an added stress of having parents or other family members with undocumented status, or being deported, it can also increase the risk of depression [32].

Some factors inherent to Latin culture can be protective for depression, family (the close connections created among Latin families that will extend the youth social support and coping strategies), religion, respect, and resilience can positively counter the effect of acculturative stress [33].

Taking into account all these factors, treatment for depression in Latinx youth should include a multimodal approach that includes a strong component of psycho-education, making sure that the parents and the youth agree and understand the importance of treatment, not including the family in the treatment of depression in youth is a big mistake since one parent might agree with mental health treatment while the other might be against, exploring and discussing the barriers for treatment is key when treating Hispanic families. Studies have reported that cognitive-behavioral therapy and interpersonal psychotherapy have demonstrated empirical support for treating depression in Latinx youth [34]. Some studies have also reported that group interventions and family interventions can be as effective as 1:1 therapy in this population [35]. Brent et al. reported Brief Behavioral Therapy (BBT) effectiveness in Hispanic youth being superior to referral to care [36]. Alegria M on a commentary of the BBT study, comments on how the lack of access to services, the delay in identifying the symptoms, and the difficulty navigating the mental health services can play an important role in the disparity of care for Latinx youth [37].

Please also see chapter on mood disorders in Hispanics in this book.

Anxiety

Anxiety disorders are the most frequent psychiatric comorbidities experienced by children and adolescents [38, 39]. Some studies have reported that Latino youth might be at high risk of developing anxiety disorders [40]. In a study of youth from Central America, authors reported a difference in the prevalence in separation anxiety disorders compared to whites, but not other differences in prevalence with other anxiety disorders [41]. Children learn to cope from observations on how their parents cope. Several studies have been reported that Latinx parents tend to worry more than their Caucasian counterparts [41] and Latinx youth reporting to worrying more than other ethnicities. This finding has been corroborated by multiple studies of Latinx youth in the USA and Latinx youth living in Latin America [42]. Age and gender seem to have an effect on the development of anxiety disorder, similar to non-Latinx white patients, younger children tend to report more fears than older children. Girls tend to report more anxiety symptoms than boys [43]. Considering the negative connotations in some Hispanic cultures about mental illness, Hispanics have used other terms that might be more socially accepted when describing anxiety, such as "nervios" or "ataque de nervios," when describing some panic attack symptoms [44, 45]. Some of the studies in Latinx youth have evaluated the use of "nervios" when describing their youth disruptive behavior and found that parents were describing some of the symptoms of ADHD, ODD, and conduct disorder as "nervios" [46].

Some studies have reported that psychosomatic symptoms might be more culturally accepted in Latinx youth than the expression of psychological symptoms of anxiety [40]. Also, studies have reported differences between Mexican American youth and American youth when reporting physical symptoms of anxiety [42]; the same was reported in Colombian youth vs. American youth [47]. When considering the evaluation of anxiety disorders in Latinx youth, it is important to remember that somatic symptoms are more prevalent and reported as a symptom of anxiety in this population, as well as functional impairment secondary to anxiety, in youth the effect of discrimination is important to take into account when examining the risk for anxiety in Latinx youth [48]. In a study of health care utilization comparing Hispanic youth vs. non-Hispanic youth, authors concluded that Hispanic youth utilized health care services at a higher rate than non-Hispanic youth and youth with an anxiety disorder had an increased number of medical visits vs. youth with no anxiety disorders [49].

Studies in Latinx youth have reported increased anxiety disorders, especially separation anxiety, especially after the change of US immigration policies in 2017 [50].

The effect of discrimination has also been studied in Mexican students. In a group of Mexican students, the level of discrimination was assessed in fifth, seventh, and ninth grade. Researchers reported that discrimination by peers in seventh and ninth grade was predictive of developing anxiety disorder by 12th grade [51, 52]. In a study evaluating data of 1466 Hispanic youth from Chicago, Miami, New York, and San Diego, authors evaluated the association between immigrant generation, acculturation, different sources of stress and resilience with outcomes related to depression, anxiety, susceptibility to alcohol consumption, and smoking and concluded the impact of acculturative stress in the development of anxiety and depression play an important role, good family function and positive ethnic identity with close community support had a protective effect in Latinx youth against the consequences of stress [53].

The Deferred Action for Childhood Arrivals (DACA) program granted temporary protection from deportation to 780,000 immigrants. In a study evaluating children's mental health issues with parents at risk of deportation, researchers reported decreased adjustment and anxiety disorders among the children of parents who were granted DACA eligibility status. DACA protection in their mothers led to improvement in their youth mental health, decreased anxiety and adjustment disorder by 4.3 percentage points which translates to a 50% decrease in youth anxiety disorders. Authors hypothesized that parents' unauthorized status is a major stressor that impacts normal child development and consequently perpetuates health inequalities transferring all the parental disadvantages to the children [54].

Like non-Hispanic whites (59%), Latinx youth were 54% more likely to have an increase in the diagnosis of anxiety and depression together [55].

Evaluating the National Comorbidity Survey adolescent supplement data on identification and referral for mental health treatment, comparing externalizing disorders (ADHD, ODD, conduct disorder) vs. internalizing disorders (MDD, Anxiety, and Dysthymia), service utilization for adolescents and ethnic differences among the services were examined. The authors concluded that identification increased the likelihood of seeking services.

Previous studies have indicated that minority parents are less likely than other parents to identify and refer their children for mental health treatment [56–58]. Even when referred, they are less likely to follow up with appropriate mental health care [59]. Some of the barriers include: perception and attitude towards mental health treatment [56], perception that treatment is not likely to help [59]; and poor knowledge about mental health conditions and their treatment [60]. Ethnic minority children are less likely to be identified and referred for mental health treatment [61]. In this study, non-Latino black students were less likely to be identified and referred for mental health services and were less likely to pursue any services [56].

In a study evaluating preferences for anxiety treatment among parents of Latinx youth and non-Latin youth, there were no differences between the two groups on parental involvement in the CBT group for youth, Latinx parents reported increased worries about the accessibility of culturally appropriate CBT that was available for their youth [62–64]. Indicated prevention and early intervention for childhood anxiety: a randomized trial with Caucasian and Hispanic/Latino youth [64]. Latinos are nevertheless less likely to seek mental health treatment [65–67].

Race and ethnicity is a strong predictor of poor adherence to treatment with antidepressant medication; in a study comparing adherence to antidepressant treatment in diverse health care settings, authors concluded that race/ethnicity was a strong predictor of early adherence, ethnic minority patients were less likely (OR 0.5–0.59) to refill their prescription than non-Hispanic whites. Patients who received their antidepressants from psychiatrist were more likely to take it vs. primary care. Patients who lived in neighborhoods with lower level of education were less likely to take their antidepressants. Patients participating in psychotherapy were more likely to refill their medication. Authors discussed the possibility that genetic differences and ethnopharmacology differences might impact the tolerability of antidepressants in minorities as one explanation of poor adherence secondary to side effects or tolerability on minorities [68].

In a study evaluating Medicaid claims data evaluating 45,816 adolescents across 9 states, authors analyzed outcomes on: minimally adequate psychotherapy (receiving at least 4 sessions in 12 weeks), minimally adequate antidepressant treatment (1 prescription for an antidepressant fill from 84–144 days), and minimally adequate treatment (medication or psychotherapy) and described that 38% received minimally adequate psychotherapy, 19% received minimally adequate pharmacotherapy, 49.9% received minimally adequate treatment, and 16% received no treatment. Percentages for Hispanic 48% and blacks 42% were significantly lower than whites 54% for minimally adequate treatment. The same was observed for blacks 20% and Hispanics 15% vs. whites 12% who received no treatment. Authors concluded that it is imperative to improve quality of depression treatment in minority youth [69].

Please also see the chapter of anxiety disorders in Hispanics in this book.

Suicide

Suicide is the tenth leading cause of death in the USA. In adolescents, it is currently the second cause of death. In the last decade, there are increased reports of suicidal ideation and behavior in Latinx, the trend being higher in Latinx youth. Studies have reported high rate of suicidal ideation among Latinx adolescents, with slightly high numbers reported recently in Latina adolescents. The rate is the highest for US-born Latinx adolescents [70].

In a study evaluating a sample from Minnesota, of Latinx from 9th–12th grade, the authors reported that 1 in 5 latinx high school students have had suicidal thoughts in the last year. In the same sample, the rate for suicidal attempt ranged from 6% to 18%. Latinas in 9–10th grade had 30–40% suicidal ideation and up to 19% reported a suicide attempt in the last 12 months [71]. When comparing Latina adolescents with their white and African American counterparts, Latina adolescents are two times more likely to report suicidal behavior [72]. Some studies have reported that latina adolescents have higher rates of depression compared to white adolescents [73].

In the National Latino and Asian American Study (NLAAS), evaluating the lifetime prevalence of suicidal ideation 10.2% and attempts 4.4% attempts authors identified female gender, acculturation, and high family conflict independent and positively associated with suicide attempts among Latinos. Latinos with a psychiatric diagnosis were more likely to report suicide attempts. Among latinx with no psychiatric disorders, being female and having high family conflict increased the risk of suicide attempts, church attendance was a protective factor for this group [22].

Among suicidal adolescents, studies have reported that up to 77% of them are not getting the mental health services they need, the problem is even worst for latinx suicidal adolescents and up to 88% are not receiving lifesaving mental health care [74]. Studies have identified some barriers and facilitators for mental health treatment among adolescents. Among the factors that will increase the likelihood of accessing services are: level of depressive symptoms, degree of suicidality, parent perception of need for treatment. Some of the factors that will enable access are: socioeconomic status, health insurance coverage, use of health services that will refer them to psychiatric care. Attitudes and perception about treatment are also an important factor. For Latinx adolescents the factors that impacted health care utilization were mood fluctuations and executive dysfunction difficulties (such as difficulty with impulsivity and attention) [75].

The role of friendship was examined in a study of 648 adolescents, half of the group were latinx. Authors make the case that having unsupportive friends is worse than having no friends at all [76, 77]. Friends was one of the strongest predictors of adolescent dysfunctional behavior [77, 78]. In a very comprehensive study of friendship and suicidality, authors concluded that social isolation and intransitive friendship were associated with SI in girls but not boys [79]. In this study, 31% of the students reported suicidal ideation, 11.5% reported non-fatal suicidal behavior in the last year. No difference in suicidality by educational status (or good academic standing). The role of friendship problems in the group of Hispanic adolescent girl accounts for 13.3% in the variance of suicidality. School disconnectedness was associated with suicidality in Hispanic girls and having delinquent friends was associated with increased suicidality with suicidality in boys and girls [80]. In a follow-up study of the same group of adolescents 3 years later, for Hispanic youth having delinquent friends predicted suicidal ideation 3 years later. For white youth having socially disconnected friends was mediated by depression as a risk factor for suicidal ideation [81].

In a study of 1651 Hispanic adolescents from 3 different states in the USA, researchers reported that suicidal ideation was higher in Hispanic female adolescents compared to Hispanic male adolescents, some of the factors that were significantly different was the impact of rupture of the family, the presence of drug abuse in the family, and decrease in family support. In Hispanic adolescent males, discrimination-related stress places an important role in the developing of suicidal ideation. Immigration stress in this study was predictive of self-harm behavior in Hispanic females [82].

Evaluating the data from the Youth Risk Behavior Surveillance System (YRBSS) cross sectional to identify latinx adolescents, authors concluded that Latina girls were more likely than boys to report suicidal ideation, they were more likely to make a suicide attempt and to report bullying and cyberbullying. Youth who carried a gun were more likely to commit suicide, boys in this group were more likely to carry guns, although in the last 10 years the number of Latino boys carrying guns have decreased [83].

In a qualitative study evaluating treatment narrative in a group of 68 Latina adolescent suicide attempters, authors described that adults in 72% of the cases initiated the referral to access for services (including family members, teachers, and counselors), up to 27% of the transfers to the hospital involved the police, 48% of the participants were referred to inpatient psychiatric services after the emergency department visit, 81% of the participants reported aggressive behaviors in conjunction with suicidal behaviors. Suicidal Latinas identified as positive experience during the admission to inpatient psychiatry: when providers frame suicidal behavior in the context of everyday experiences, when the treatment team decreased the stigma associated with suicidal behavior. In the outpatient services they found helpful when the provider: normalized the experience, fostering agency and improve family interactions. Clinician strategies on communication were key to engage this population, patients identified that communicating their distress had cathartic effect, the more they communicated the more they felt connected with the clinician, inviting teens to make decisions on choosing their treatment help later adherence to their treatment [84].

In a study evaluating data from LGBT latinx youth from 2017-National Youth Risk Behavioral Survey, 451 participants who self-identified as Latinx LGBT data was analyzed, suicidal ideation was reported in 40% of the participants, patients reported having a plan in 34% of the cases, and a suicide attempt in 21% of the cases. Suicide attempts were significantly associated with cannabis use, sexual assault, history of depression, and being bullied at school [85].

Attention Deficit Hyperactivity Disorder (ADHD)

ADHD and learning disorders are the most common neurodevelopmental disorders of childhood [86]. In 2016–2018, 13.8% of children aged 3–17 suffered ADHD or a learning disability [87]. Data from 2009–2017 National Health Interview Survey found a significant increasing trend in the prevalence of attention-deficit/hyperactivity disorder in the USA, ranging between 8.5% and 9.5% [88]. The analysis of National Survey of Children's Health from the year 2003, 2007, and 2011 found that although the prevalence of ADHD remained highest among the white population, the most notable increase in prevalence was noted among the Latino population, which increased by 83% from 2003 to 2011, there is an 83% increase in ADHD [89].

The Latino population is deemed particularly vulnerable to the lack of recognition of ADHD because of barriers of various reasons. These reasons include lack of culturally sensitive of diagnostic instrumentation, language barrier, or difficult therapeutic relationship between patients and clinicians [90, 91]. Knowledge about ADHD and its causes also appears limited in the Latino population, as Latinos are less likely than Caucasians to report familiarity with ADHD in large national surveys [91]. In addition to these general factors, one must consider socioeconomic influences to problem recognition in Latino youth, as Latino families are disproportionately impoverished and underinsured in the USA. The lack of parental employment-linked benefits, procedural barriers to enrollment, and lack of clarification on eligibility for children of noncitizen parents are associated with low levels of insurance coverage among Latino children [92, 93]. Latino youth appear to receive fewer health-care and follow-up visits with physicians than Caucasian youth, thus providing fewer opportunities for health care professionals to inquire about ADHD problems [94]. Additionally, there are lower rates of parent–teacher communication and family-school involvement in Latinos than in non-Latino communities [95]. The discussion of child behavior between teachers and parents may further suffer from language barriers.

Latino families lack knowledge about etiology and effective treatment for ADHD. Many Latino families believe the cause of ADHD to be family-related factors, such as lack of parental attention or discipline, and believe that the symptoms are likely to go away on their own [96]. Additionally, Latino parents perceive children's behavior as an issue with lack of manners or disrespect for authority rather than accepting their child's behavior problems as a medical disability [97]. In contrast, Latino parents may be less concerned about ADHD symptomology, particularly in the hyperactivity-impulsivity domain. They consider it a saturation point compared to problem recognition, and instead may believe these behaviors are a regular aspect of child development [98, 99]. Mothers typically acknowledge the problem after their children enter school programs and school reports negative behavior. Orientation to cultural values may help explain Latino parents' beliefs about the cause and burden of various ADHD behaviors. Collectivistic values, such as *personalismo* (i.e., emphasis on warm, interpersonal closeness), *familismo* (i.e., emphasis on family connectedness and loyalty), *respeto* (i.e., emphasis on obedience and compliance to authority), *spiritualismo* (i.e., religiosity), and traditional gender roles or *machismo* (i.e., emphasis on masculine pride and feminine submissiveness) are particularly relevant to perceptions about and treatment of child's behavior problems [91, 100–102].

Accurate assessment and diagnosis of ADHD relies heavily on parental report, mainly when using tools like standardized behavior rating scales. Studies suggest that Latino youth were less likely than Caucasian and African American children to be diagnosed with ADHD, particularly due to parent-reported symptoms [103, 104]. Conversely, once diagnosed with ADHD, there is an increase in utilization of mental health services among Latino children. Having an ADHD diagnosis significantly predicted mental health service use across multiple sectors of care. A large study examining mental health service use for ADHD during a 3-year period among a sample of 2491 Puerto Rican children living in the South Bronx, NY ($n = 1138$) and Puerto Rico ($n = 1353$) found that Puerto Rican children with ADHD were about three times more likely to use mental health services than children diagnosed with a different mental health disorder [105]. This study also indicated that although prevalence rates of ADHD were similar across the two contexts (4%–7%), general mental health service use was significantly lower among children living in Puerto Rico. This difference is likely partly related to a greater availability of mental health services in New York than Puerto Rico.

Therefore, the literature overall suggests that Latino children are less likely to be diagnosed with ADHD compared to other racial populations but once the diagnosis is confirmed, there is robust adherence with treatment among them. These findings have important implications concerning the assessment of ADHD among Latino children. They highlight the need to consider factors such as parental perceptions of child behavior in the assessment process. In addition to contributing to mental health service's underutilization, language serves as a barrier to advancing research about this underserved population. Both for convenience and practicality, many epidemiological and treatment outcome studies screen for and exclude all individuals who do not speak English, which has resulted in a lack of research examining prevalence rates and treatment outcomes for Latino youth [106–108]. The studies overcome language barrier report that Latino youth are at a higher risk for several mental health problems [109, 110], including behavioral disorders, such as ADHD [106, 109].

Given the frequency and long-lasting impact of ADHD [110–112], and the higher risk for behavioral disorders in Latino youth, it seems particularly important to devote more research efforts and resources to examining ADHD in Latinos.

Some important considerations to take into the account before providing mental health services for Latinx youth are (adapted from 11*):*

1. It is important to address the English language proficiency in the family. If the youth or the family need translator services, the use of medical interpreters should be essential; using the youth or other family members as translators should be discouraged.

   Discuss with the interpreter before the appointment the importance of fully translating all the words to provider for a better understanding of whole family.

Consider that provider might need extra time for a first-time patient if interpreter services are being used.

2. Be prepared for the first visit (think Latinx families will potentially bring many family members to the appointment), have appropriate seating for everybody in the family in your office. For Latinx families it is very important to build a good relationship based on trust, meet each one of the members of the families who came to the appointment. Give each one of them the opportunity to participate at the beginning of the appointment, then ask for permission to talk to the youth along (to improve your relationship with the patient), then bring the family back at the end of the appointment.

   For Latinx families, it is always important to explore the level of connectedness, their belief system, and the barriers against mental health care, identify if there are different views about treatment in the family, and make sure to engage them on the decision to improve treatment adherence.
3. Always listen to each family's unique history, take time to understand their immigration status, put them at ease in the office, and understand how their migratory experiences can impact the family and youth's mental health. Always ask about country of origin, explore exposure to past or current violence experiences. Explore how many people or families live under the same roof and how they are currently getting along, exploring challenges related to intergenerational conflict, exploring deportation fears and past history of intergenerational trauma.
4. Explore how the family is integrating into the American culture and the role that acculturative stress is playing in the mental well-being of the child. Explore gender roles and gender perception in the family (with the parents and when they leave the room with the youth).
5. Explore how the family has experienced mental illness in the past. Their beliefs around mental illness and mental health treatment are clear about their expectations on the doctor–patient relationship. Explore if stigma played a role in accessing or delaying mental health care. Explore if they consulted other healers and the recommendations for the current issues, explore any herbal supplements or other alternative treatment that the patient might be taking or is considering to take, and do psychoeducation about the potential interactions between them some of the herbal remedies and antidepressants.
6. Organizations like NAMI, can increase resilience, encourage activities and social support. Provide written documents (ask the family if they prefer them in Spanish or English) about the specific mental health issues that the youth is facing and the treatment indicated for that specific condition. Encourage them to learn about the illness, use appropriate sources, ask questions and communicate if they are having side effects from their medication or considering stopping treatment early.
7. Decreasing stigma is probably the best intervention to increase treatment adherence, focus on psychoeducation in each appointment, and reassess family, and youth commitment to continue treatment at each appointment.

References

1. Mota NB, Copelli M, Ribeiro S. Computational tracking of mental health in youth: Latin American contributions to a low-cost and effective solution for early psychiatric diagnosis. New Dir Child Adolesc Dev. 2016 Jun;2016(152):59–69. https://doi.org/10.1002/cad.20159.
2. Fleitlich BW, Goodman R. Epidemiologia. Rev Bras Psiquiatr. 2000;22(Suppl. 2):2–6.
3. Roberts RE, Attkisson CC, Rosenblatt A. Prevalence of psychopathology among children and adolescents. Am J Psychiatry. 1998 Jun;155(6):715–25. https://doi.org/10.1176/ajp.155.6.715.
4. De la Barra MF, Vicente PB, Saldivia BS, Melipillán AR. Estudio de epidemiología psiquiátrica en niños y adolescentes en Chile. Estado actual [Investigation of psychiatric epidemiology among children and adolescents in Chile]. Revista Médica Clínica Las Condes. 2012;23(5):521–9.
5. Ford T, Goodman R, Meltzer H. The British child and adolescent mental health survey 1999: the prevalence of DSM-IV disorders. J Am Acad Child Adolesc Psychiatry. 2003 Oct;42(10):1203–11. https://doi.org/10.1097/00004583-200310000-00011.
6. Belfer ML, Rohde LA. Child and adolescent mental health in Latin America and the Caribbean: problems, progress, and policy research. Rev Panam Salud Publica. 2005;18(4–5):359–65. https://doi.org/10.1590/s1020-49892005000900016.
7. Oschilewsky RC, Gómez CM, Belfort E. Child psychiatry and mental health in Latin America. Int Rev Psychiatry. 2010;22(4):355–62. https://doi.org/10.3109/09540261.2010.503692.
8. Rohde LA, Celia S, Berganza C. Systems of care in South America. In: Remschmidt H, Belfer M, Gooyer I, editors. Facilitating pathways, care, treatment and prevention in child and adolescent mental health; 2004. p. 42–51.
9. Duarte C, Hoven C, Berganza C, Bordin I, Bird H, Miranda CT. Child mental health in Latin America: present and future epidemiologic research. Int J Psychiatry Med. 2003;33(3):203–22. https://doi.org/10.2190/4WJB-BW16-2TGE-565W.
10. Betancourt TS, Newnham EA, Birman D, Lee R, Ellis BH, Layne CM. Comparing trauma exposure, mental health needs, and service utilization across clinical samples of refugee, immigrant, and U.S.-origin children. J Trauma Stress. 2017;30(3):209–18. https://doi.org/10.1002/jts.22186. Epub 2017 Jun 6
11. Concepcion Zayas MT, Fortuna LR, Cullins LM. Depression in Latino and immigrant refugee youth: clinical opportunities and considerations. Child Adolesc Psychiatr Clin N Am. 2019;28(3):483–95. https://doi.org/10.1016/j.chc.2019.02.013. Epub 2019 Apr 4
12. Latino children are the fastest growing population in the US. https://www-census-gov.ccmain.ohionet.org/newsroom/facts-for-features/2018/hispanic-heritage-month.html
13. Foxen P, Mather M, Foxen P, et al. Towards a more equitable future the trends and challenges facing America's Latino children population. Washington D.C.: National Council of La Raza; 2016. p. 9.
14. Alegria M, Vallas M, Pumariega AJ. Racial and ethnic disparities in pediatric mental health. Child Adolesc Psychiatr Clin N Am. 2010 Oct;19(4):759–74. https://doi.org/10.1016/j.chc.2010.07.001.
15. Potochnick SR, Perreira KM. Depression and anxiety among first-generation immigrant Latino youth: key correlates and implications for future research. J Nerv Ment Dis. 2010 Jul;198(7):470–7. https://doi.org/10.1097/NMD.0b013e3181e4ce24.
16. Rodríguez J, De La Torre A, Miranda CT. Mental health in situations of armed conflict. Biomedica. 2002;22:337–46.
17. Fazel M, Reed RV, Panter-Brick C, Stein A. Mental health of displaced and refugee children resettled in high-income countries: risk and protective factors. Lancet. 2012;379(9812):266–82. https://doi.org/10.1016/S0140-6736(11)60051-2. Epub 2011 Aug 9
18. Lê F, Tracy M, Norris FH, Galea S. Displacement, county social cohesion, and depression after a large-scale traumatic event. Soc Psychiatry Psychiatr Epidemiol. 2013;48(11):1729–41. https://doi.org/10.1007/s00127-013-0698-7. Epub 2013 May 5

19. Thapar A, Collishaw S, Pine DS, Thapar AK. Depression in adolescence. Lancet. 2012;379(9820):1056–67. https://doi.org/10.1016/S0140-6736(11)60871-4. Epub 2012 Feb 2
20. Kann L, McManus T, Harris WA, Shanklin SL, Flint KH, Queen B, Lowry R, Chyen D, Whittle L, Thornton J, Lim C, Bradford D, Yamakawa Y, Leon M, Brener N, Ethier KA. Youth risk behavior surveillance–United States, 2017. MMWR Surveill Summ. 2018 Jun 15;67(8):1–114. https://doi.org/10.15585/mmwr.ss6708a1.
21. Guzmán A, Koons A, Postolache TT. Suicidal behavior in Latinos: focus on the youth. Int J Adolesc Med Health. 2009;21(4):431–9. https://doi.org/10.1515/ijamh.2009.21.4.431.
22. Fortuna LR, Perez DJ, Canino G, Sribney W, Alegria M. Prevalence and correlates of lifetime suicidal ideation and suicide attempts among Latino subgroups in the United States. J Clin Psychiatry. 2007 Apr;68(4):572–81. https://doi.org/10.4088/jcp.v68n0413.
23. https://www.samhsa.gov/sites/default/files/covid19-behavioral-health-disparities-black-latino-communities.pdf.
24. Marrast L, Himmelstein DU, Woolhandler S. Racial and ethnic disparities in mental health care for children and young adults: a national study. Int J Health Serv. 2016;46(4):810–24. https://doi.org/10.1177/0020731416662736. Epub 2016 Aug 12
25. Olfson M, He JP, Merikangas KR. Psychotropic medication treatment of adolescents: results from the national comorbidity survey-adolescent supplement. J Am Acad Child Adolesc Psychiatry. 2013;52(4):378–88. https://doi.org/10.1016/j.jaac.2012.12.006. Epub 2013 Mar 7
26. Kirby JB, Hudson J, Miller GE. Explaining racial and ethnic differences in antidepressant use among adolescents. Med Care Res Rev. 2010;67:342–63.
27. Goodman E, Slap GB, Huang B. The public health impact of socioeconomic status on adolescent depression and obesity. Am J Public Health. 2003 Nov;93(11):1844–50. https://doi.org/10.2105/ajph.93.11.1844.
28. https://www.census.gov/content/dam/Census/library/publications/2015/demo/p60-252.pdf
29. Ornelas IJ, Perreira KM. The role of migration in the development of depressive symptoms among Latino immigrant parents in the USA. Soc Sci Med. 2011;73(8):1169–77. https://doi.org/10.1016/j.socscimed.2011.07.002. Epub 2011 Jul 29
30. Kam JA, Lazarevic V. The stressful (and not so stressful) nature of language brokering: identifying when brokering functions as a cultural stressor for Latino immigrant children in early adolescence. J Youth Adolesc. 2014;43(12):1994–2011. https://doi.org/10.1007/s10964-013-0061-z. Epub 2013 Nov 17
31. Benner AD, Graham S. Latino adolescents' experiences of discrimination across the first 2 years of high school: correlates and influences on educational outcomes. Child Dev. 2011;82(2):508–19. https://doi.org/10.1111/j.1467-8624.2010.01524.x. Epub 2011 Mar 9
32. Arbona C, Olvera N, Rodriguez N, et. al. Acculturative stress among documented and undocumented Latino immigrants in the United States. Hisp J Behav Sci. 2010;32:362–84.
33. Lorenzo-Blanco EI, Unger JB, Baezconde-Garbanati L, et. al. Acculturation, enculturation, and symptoms of depression in Hispanic youth: the roles of gender, Hispanic cultural values, and family functioning. J Youth Adolesc. 2012;41:1350–65.
34. Collado A, Lim AC, MacPherson L. A systematic review of depression psychotherapies among Latinos. Clin Psychol Rev. 2016;45:193–209. https://doi.org/10.1016/j.cpr.2016.04.001. Epub 2016 Apr 9
35. Ingoldsby EM. Review of interventions to improve family engagement and retention in parent and child mental health programs. J Child Fam Stud. 2010 Oct 1;19(5):629–45. https://doi.org/10.1007/s10826-009-9350-2.
36. Brent DA, Porta G, Rozenman MS, Gonzalez A, Schwartz KTG, Lynch FL, Dickerson JF, Iyengar S, Weersing VR. Brief behavioral therapy for pediatric anxiety and depression in primary care: a follow-up. J Am Acad Child Adolesc Psychiatry. 2020;59(7):856–67. https://doi.org/10.1016/j.jaac.2019.06.009. Epub 2019 Jul 3
37. Alegría M. Editorial: increasing access to mental health Care for Latinx Children: moving in the right direction. J Am Acad Child Adolesc Psychiatry. 2020;59(7):800–2. https://doi.org/10.1016/j.jaac.2019.10.006. Epub 2019 Oct 22

38. Berstein GA, Borchardt CM. Anxiety disorders of childhood and adolescence: a critical review. J Am Acad Child Adolesc Psychiatry. 1991;30:519–32.
39. Kashani JH, Orvaschel H. A community study of anxiety in children and adolescents. Am J Psychiatry. 1990;147:313–8.
40. Varela RE, et al. Parenting practices, interpretive biases, and anxiety in Latino children. J Anxiety Disord. 2013;27(2):171–7.
41. Ginsburg GS, Silverman WK. Phobic disorders in Hispanic and European-American youth. J Anxiety Disord. 1996;10:517–28.
42. Varela RE, Sanchez-Sosa JJ, Biggs BK, Luis TM. Anxiety symptoms and fears in Hispanic and European American children: cross-cultural measurement equivalence. J Psychopathol Behav Assess. 2008;30:132–45.
43. Weems CF, Costa NM. Developmental differences in the expression of childhood anxiety symptoms and fears. J Am Acad Child Adolesc Psychiatry. 2005;44:656–63.
44. Jenkins JH. Ethnopsychiatric interpretations of schizophrenic illness: The problem of nervios within Mexican- American families. Cult Med Psychiatry. 1988a;12:301–29.
45. Milstein G, Guarnaccia P, Midlarsky E. Ethnic differences in the interpretation of mental illness: perspectives of caregivers. Res Community Ment Health. 1995;8:155–78.
46. Arcia E, Castillo H, Fernandez MC. Maternal cognitions about distress and anxiety in young Latino children with disruptive behaviors. Transcult Psychiatry. 2004;41:99–119.
47. Varela RE, Weems CF, Berman SL, Hensley L, Rodriguez de Bernal MC. Internalizing symptoms in Latinos: The role of anxiety sensitivity. J Youth Adolesc. 2007;36:429–40.
48. Escovar EL, Craske M, Roy-Byrne P, Stein MB, Sullivan G, Sherbourne CD, Bystritsky A, Chavira DA. Cultural influences on mental health symptoms in a primary care sample of Latinx patients. J Anxiety Disord. 2018;55:39–47. https://doi.org/10.1016/j.janxdis.2018.03.005. Epub 2018 Mar 12
49. Bilsky SA, Dueweke AR, Bridges AJ. The impact of anxiety on health care utilization among Hispanic and non-Hispanic youth. J Health Psychol. 2020;25(7):986–92. https://doi.org/10.1177/1359105317740738. Epub 2017 Nov 10
50. Brown CR, Shramko M, Garcia-Huidobro D, Miller KK, Ogugua PBF, Svetaz MV. Increased rates of parental separation and anxiety among Latinx youth children and youth services review, vol. 116; 2020. p. 105225.
51. Stein GL, Castro-Schilo L, Cavanaugh AM, Mejia Y, Christophe NK, Robins R. When discrimination hurts: The longitudinal impact of increases in peer discrimination on anxiety and depressive symptoms in Mexican-origin youth. J Youth Adolesc. 2019;48(5):864–75. https://doi.org/10.1007/s10964-019-01012-3. Epub 2019 Mar 16
52. Varela RE, Hensley-Maloney L. The influence of culture on anxiety in Latino youth: a review. Clin Child Fam Psychol Rev. 2009 Sep;12(3):217–33. https://doi.org/10.1007/s10567-009-0044-5.
53. Perreira KM, Marchante AN, Schwartz SJ, Isasi CR, Carnethon MR, Corliss HL, Kaplan RC, Santisteban DA, Vidot DC, Van Horn L, Delamater AM. Stress and resilience: key correlates of mental health and substance use in the Hispanic community health study of Latino youth. J Immigr Minor Health. 2019 Feb;21(1):4–13. https://doi.org/10.1007/s10903-018-0724-7.
54. Hainmueller J, Lawrence D, Martén L, Black B, Figueroa L, Hotard M, Jiménez TR, Mendoza F, Rodriguez MI, Swartz JJ, Laitin DD. Protecting unauthorized immigrant mothers improves their children's mental health. Science. 2017;357(6355):1041–4. https://doi.org/10.1126/science.aan5893. Epub 2017 Aug 31
55. Weller BE, Blanford KL, Butler AM. Estimated prevalence of psychiatric comorbidities in U.S. adolescents with depression by race/ethnicity, 2011-2012. J Adolesc Health. 2018 Jun;62(6):716–21. https://doi.org/10.1016/j.jadohealth.2017.12.020.
56. Alegría M, Lin JY, Green JG, Sampson NA, Gruber MJ, Kessler RC. Role of referrals in mental health service disparities for racial and ethnic minority youth. J Am Acad Child Adolesc Psychiatry. 2012;51(7):703–711.e2. https://doi.org/10.1016/j.jaac.2012.05.005. Epub 2012 Jun 1

57. Wachtel J, Rodrigue JR, Geffken GR, Graham-Pole J, Turner C. Children awaiting invasive medical procedures: do children and their mothers agree on Child's level of anxiety? J Pediatr Psychol. 1994;19(6):723–35.
58. Walton JW, Johnson SB, Algina J. Mother and child perceptions of child anxiety: effects of race, health status, and stress. J Pediatr Psychol. 1999;24(1):29–39.
59. McMiller WP, Weisz JR. Help-seeking preceding mental health clinic intake among African-American, Latino, and Caucasian youths. J Am Acad Child Adolesc Psychiatry. 1996 Aug;35(8):1086–94. https://doi.org/10.1097/00004583-199608000-00020.
60. Langrehr KJ. Racial distinctions in the psychosocial histories of incarcerated youth. Psychol Serv. 2011;8(1):23–35.
61. Wood PA, Yeh M, Pan D, et al. Exploring the relationship between race/ethnicity, age of first school-based services utilization, and age of first specialty mental health Care for at-Risk Youth. Ment Health Serv Res. 2005;7:185–96.
62. Seligman LD, Hovey JD, Ibarra M, Hurtado G, Marin CE, Silverman WK. Latino and non-Latino parental treatment preferences for child and adolescent anxiety disorders. Child Psychiatry Hum Dev. 2020;51:617–24.
63. Pina AA, Silverman WK, Fuentes RM, Kurtines WM, Weems CF. Exposure-based cognitive-behavioral treatment for phobic and anxiety disorders: treatment effects and maintenance for Hispanic/Latino relative to European-American youths. J Am Acad Child Adolesc Psychiatry. 2003 Oct;42(10):1179–87. https://doi.org/10.1097/00004583-200310000-00008.
64. Pina AA, Zerr AA, Villalta IK, Gonzales NA. Indicated prevention and early intervention for childhood anxiety: a randomized trial with Caucasian and Hispanic/Latino youth. J Consult Clin Psychol. 2012;80(5):940–6. https://doi.org/10.1037/a0029460. Epub 2012 Jul 23
65. Cabassa LJ, Zayas LH, Hansen MC. Latino adults' access to mental health care: a review of epidemiological studies. Admin Pol Ment Health. 2006 May;33(3):316–30. https://doi.org/10.1007/s10488-006-0040-8.
66. Cardemil EV, Adams ST, Calista JL, Connell J, Encarnación J, Esparza NK, Frohock J, Hicks E, Kim S, Kokernak G, McGrenra M, Mestre R, Pérez M, Pinedo TM, Quagan R, Rivera C, Taucer P, Wang E. The Latino mental health project: a local mental health needs assessment. Admin Pol Ment Health. 2007;34(4):331–41. https://doi.org/10.1007/s10488-007-0113-3. Epub 2007 Feb 6
67. Kouyoumdjian H, Zamboanga BL, Hansen DJ. Barriers to community mental health services for Latinos: treatment considerations. Clin Psychol Sci Pract. 2003;10(4):394–422.
68. Rossom RC, Shortreed S, Coleman KJ, Beck A, Waitzfelder BE, Stewart C, Ahmedani BK, Zeber JE, Simon GE. Antidepressant adherence across diverse populations and healthcare settings. Depress Anxiety. 2016;33(8):765–74. https://doi.org/10.1002/da.22532. Epub 2016 Jun 20
69. Cummings JR, Ji X, Lally C, Druss BG. Racial and ethnic differences in minimally adequate depression care among Medicaid-enrolled youth. J Am Acad Child Adolesc Psychiatry. 2019;58(1):128–38. https://doi.org/10.1016/j.jaac.2018.04.025. Epub 2018 Oct 17
70. Silva C, Van Orden KA. Suicide among Hispanics in the United States. Curr Opin Psychol. 2018;22:44–9. https://doi.org/10.1016/j.copsyc.2017.07.013. Epub 2017 Jul 18
71. Garcia C, Skay C, Sieving R, Naughton S, Bearinger LH. Family and racial factors associated with suicide and emotional distress among latino students. J Sch Health. 2008 Sep;78(9):487–95. https://doi.org/10.1111/j.1746-1561.2008.00334.x.
72. https://www.cdc.gov/mmwr/preview/mmwrhtml/ss5704a1.htm
73. Eaton DK, Foti K, Brener ND, Crosby AE, Flores G, Kann L. Associations between risk behaviors and suicidal ideation and suicide attempts: do racial/ethnic variations in associations account for increased risk of suicidal behaviors among Hispanic/Latina 9th- to 12th-grade female students? Arch Suicide Res. 2011;15(2):113–26. https://doi.org/10.1080/13811118.2011.565268.
74. Marshall GN, Berthold SM, Schell TL, Elliott MN, Chun CA, Hambarsoomians K. Rates and correlates of seeking mental health services among Cambodian refugees. Am J Public Health. 2006 Oct;96(10):1829–35. https://doi.org/10.2105/AJPH.2006.086736.

75. Alonzo D, Conway A, Modrek AS. Latino suicidal adolescent psychosocial service utilization: The role of mood fluctuations and inattention. J Affect Disord. 2016;190:616–22. https://doi.org/10.1016/j.jad.2015.10.036. Epub 2015 Oct 28
76. Güroğlu B, Van Lieshout CFM, Haselager GJT, Scholte RHJ. Similarity and complementarity of behavioral profiles of friendship types and types of friends: friendships and psychosocial adjustment. J Res Adolesc. 2007;17:357–86.
77. Prinstein JJ, Boergers J, Spirito A. Adolescents' and their friends' health-risk behavior: factors that alter or add to peer influence. J Pediatr Psychol. 2001;26:287–98.
78. Tani CR, Chavez EL, Deffenbacher JL. Peer isolation and drug use among white, non-Hispanic and Mexican-American adolescents. Adolescence. 2001;36:127–39.
79. Bearman PS, Moody J. Suicide and friendships among American adolescents. Am J Public Health. 2004 Jan;94(1):89–95. https://doi.org/10.2105/ajph.94.1.89.
80. Winterrowd E, Canetto SS, Chavez EL. Friendship factors and suicidality: common and unique patterns in Mexican American and European American youth. Suicide Life Threat Behav. 2011;41(1):50–65. https://doi.org/10.1111/j.1943-278X.2010.00001.x. Epub 2011 Jan 19
81. Winterrowd E, Canetto SS. The long-lasting impact of adolescents' deviant friends on suicidality: a 3-year follow-up perspective. Soc Psychiatry Psychiatr Epidemiol. 2013;48(2):245–55. https://doi.org/10.1007/s00127-012-0529-2. Epub 2012 Jun 21
82. Cervantes RC, Goldbach JT, Varela A, Santisteban DA. Self-harm among Hispanic adolescents: investigating the role of culture-related stressors. J Adolesc Health. 2014;55(5):633–9. https://doi.org/10.1016/j.jadohealth.2014.05.017. Epub 2014 Jul 29
83. Romero AJ, Bauman S, Borgstrom M, Kim SE. Examining suicidality, bullying, and gun carrying among Latina/o youth over 10 years. Am J Orthopsychiatry. 2018;88(4):450–61. https://doi.org/10.1037/ort0000323.
84. Hausmann-Stabile C, Gulbas L, Zayas LH. Treatment narratives of suicidal Latina teens. Arch Suicide Res. 2018;22(1):165–72. https://doi.org/10.1080/13811118.2017.1304305. Epub 2017 Apr 7
85. Boyas JF, Villarreal-Otálora T, Alvarez-Hernandez LR, Fatehi M. Suicide ideation, planning, and attempts: the case of the Latinx LGB youth. Health Promot Perspect. 2019 Aug 6;9(3):198–206. https://doi.org/10.15171/hpp.2019.28.
86. American Psychiatric Association. Diagnostic and statistical manual of mental disorders. 5th ed. Washington, D.C: American Psychiatric Publishing; 2013. p. DSM-5. 5.
87. https://www.cdc.gov/nchs/products/databriefs/db358.htm#ref1
88. Zablotsky B, Black LI, Maenner MJ, Schieve LA, Danielson ML, Bitsko RH, et al. Prevalence and trends of developmental disabilities among children in the United States: 2009–2017. Pediatrics. 2019;144(4):e20190811.
89. Collins KP, Cleary SD. Racial and ethnic disparities in parent-reported diagnosis of ADHD: National Survey of Children's health (2003, 2007, and 2011). J Clin Psychiatry. 2016 Jan;77(1):52–9. https://doi.org/10.4088/JCP.14m09364.
90. Haack LM, Gerdes AC, Lawton KE, Schneider BW. Understanding and measuring functional impairment in diverse children with ADHD: development of the ADHD-FX scale with an at-risk. Community Sample J Atten Disord. 2016;20(6):487–500. https://doi.org/10.1177/1087054714527791. Epub 2014 Apr 2
91. Rothe EM. Considering cultural diversity in the management of ADHD in Hispanic patients. J Natl Med Assoc. 2005 Oct;97(10 Suppl):17S–23S.
92. DeNavas-Walt C, Proctor BD, Smith JC. Income, poverty, and health insurance coverage in the United States: 2009. U.S. Census Bureau; 2009.
93. Zambrana RE, Carter-Pokras O. Improving health insurance coverage for Latino children: a review of barriers, challenges and state strategies. J Natl Med Assoc. 2004 Apr;96(4): 508–23.
94. Stevens J, Harman JS, Kelleher KJ. Race/ethnicity and insurance status as factors associated with ADHD treatment patterns. J Child Adolesc Psychopharmacol. 2005 Feb;15(1):88–96. https://doi.org/10.1089/cap.2005.15.88.

95. Wong SW, Hughes JN. Ethnicity and language contributions to dimensions of parent involvement. School Psych Rev. 2006;35(4):645–62.
96. Gerdes AC, Lawton KE, Haack LM, Schneider BW. Latino parental help seeking for childhood ADHD. Admin Pol Ment Health. 2014 Jul;41(4):503–13. https://doi.org/10.1007/s10488-013-0487-3.
97. Perry CE, Hatton D, Kendall J. Latino parents' accounts of attention deficit hyperactivity disorder. J Transcult Nurs. 2005 Oct;16(4):312–21. https://doi.org/10.1177/1043659605278938.
98. Arcia E, Fernandez MC. From awareness to acknowledgement: the development of concern among Latina mothers of children with disruptive behaviors. J Atten Disord. 2003 Jun;6(4):163–75.
99. Gerdes AC, Lawton KE, Haack LM, Hurtado GD. Assessing ADHD in Latino families: evidence for moving beyond symptomatology. J Atten Disord. 2013 Feb;17(2):128–40.
100. Alvarado C, Modesto-Lowe V. Improving treatment in minority children with attention deficit/hyperactivity disorder. Clin Pediatr (Phila). 2017;56(2):171–6. https://doi.org/10.1177/0009922816645517. Epub 2016 Jul 20
101. Barker CH, Cook KL, Borrego J. Jr addressing cultural variables in parent training programs with Latino families. Cogn Behav Pract. 2010;17(2):157–66.
102. Lawton KE, Gerdes AC, Haack LM, Schneider B. Acculturation, cultural values, and Latino parental beliefs about the etiology of ADHD. Admin Pol Ment Health. 2014 Mar;41(2):189–204. https://doi.org/10.1007/s10488-012-0447-3.
103. Leslie LK, Lambros KM, Aarons GA, et al. School-based service use by youth with ADHD in public-sector settings. J Emot Behav Disord. 2008;16:163–77.
104. Pastor PN, Reuben CA. Diagnosed attention deficit hyperactivity disorder and learning disability: United States, 2004–2006. Vital Health Stat. 2008 Jul;10(237):1–14.
105. Bird HR, Shrout PE, Duarte CS, Shen S, Bauermeister JJ, Canino G. Longitudinal mental health service and medication use for ADHD among Puerto Rican youth in two contexts. J Am Acad Child Adolesc Psychiatry. 2008 Aug;47(8):879–89. https://doi.org/10.1097/CHI.0b013e318179963c.
106. Flores G, Guentes-Afflick E, Barbot O, Carter-Pokras O, Claudio L, Lara M, et al. The health of Latino children: urgent priorities, unanswered questions, and a research agenda. J Am Med Assoc. 2002;288:82–90.
107. Callejas L, Nesman R, Mowery D, Garnache P. Research with Latino populations. In: Hernandez M, et al., editors. Examining the research base supporting culturally competent children's mental health services. Tampa: Louis de la Parte Florida Mental Health Institute; 2006. p. 83–116.
108. Peifer KL, The-wei H, Vega W. Help seeking by persons of Mexican origin with functional impairments. Psychiatr Serv. 2000;51:1293–8.
109. U.S. Department of Health and Human Services. Mental health: culture, race, and ethnicity–a supplement to mental health: a report of the surgeon general. Rockville: U.S. Department of Health and Human Services, Substance Abuse and Mental Health Services Administration, Center for Mental Health Services; 2001.
110. American Psychiatric Association. Diagnostic and statistical Manuel of mental disorders, fourth edition, text revision (DSMIV-TR). Washington, DC: American Psychiatric Association; 2000.
111. Smith BH, Barkley RA, Shapiro CJ. Attention deficit/hyperactivity disorder. In: Mash EJ, Barkley RA, editors. Treatment of childhood disorders. 3rd ed. New York: Guildford; 2006. p. 65–136.
112. Pelham WE. Attention deficit hyperactivity disorder: diagnosis, assessment, nature, etiology, and treatment. Unpublished manuscript, State University of New York at Buffalo. 2002.

Cultural Sensitive Services for Latinos

12

Andrea Mendiola-Iparraguirre, Juan Rodriguez-Guzman, and Esperanza Diaz

Introduction

Healthcare problems, including mental health issues, are exacerbated because of ethnic health disparities [1–3]. For instance, an analysis of more than 45 years of data found that Latinos have lower access to specialty mental healthcare [4], and these numbers prevail despite adjustments to socioeconomic status [5]. Factors that could explain the significant disparity in healthcare are language fluency, cultural differences, access to Medicaid specialty services in Latino neighborhoods, differences in recognition of mental health problems, and lower quality of mental healthcare [5]. Additionally, social determinants of health impact access to care, appropriate treatment, equality in treatment, follow-up and continuing care services, and recovery from mental illness [6, 7].

Therefore, developing culturally sensitive mental health services is essential to provide appropriate care and to mitigate these disparities.

In this chapter, we summarize the recovery-oriented and culturally sensitive services for Latinos, including services provided in main facilities, community organizations and peer support, their benefits and challenges, and suggestions to implement them.

A. Mendiola-Iparraguirre (✉) · J. Rodriguez-Guzman · E. Diaz
Department of Psychiatry, Yale University, New Haven, CT, USA
e-mail: andrea.mendiola@yale.edu; juan.rodriguezguzman@yale.edu; esperanza.diaz@yale.edu

R. Castilla-Puentes, T. Falcone (eds.), *Mental Health for Hispanic Communities*,
https://doi.org/10.1007/978-3-031-13195-0_12

Cultural Sensitive Services for Latinos in the United States over Time

The development of Hispanic mental health services needed multiple perspectives: Psychiatry, Cultural Psychiatry, Psychology, Social Work, Anthropology, Sociology, Epidemiology, Public Health, among others. Professionals from these disciplines have worked together to provide understandings, develop theories, research, and interviewing skills. The initial aim of the researchers focused on describing the different perspectives a culture could provide in mental illness and health beliefs according to religion, spirituality, terms used by the population, language, stigma of mental illness, and indigenous treatments [8–10].

In 1995, the DSM published a glossary of terms used by different cultures to talk about mental illness. For Hispanics, the examples include *susto, nervios, mal de ojo, ataque de nervios* [11]. They emphasize the importance of understanding a person's culture in order to be culturally sensitive. They provide a way to name presentations outside of the DSM psychiatric diagnoses. Along with the glossary of cultural-bound syndromes, the DSM published the Cultural Formulation outline, which is useful up to this day [12]. It is the basis for the Cultural Formulation Interview (CFI), developed and published in DSM 5 [13]. The CFI provides sixteen questions to elicit information about the person to understand the unique perspective and wishes for treatment and to develop an individualized formulation avoiding stereotypes.

The influences on Hispanic mental health outcomes were multiple: disparities, immigration, the differences between first, second and third generations of Hispanics, the diversity of Latinos, the Latino Paradox, discrimination, and diagnoses all based on studies made in America for Americans. It was noticed that the expression of emotions takes many ways, and for Hispanics, somatic symptoms were prevalent [14]. Somatization takes many Hispanic to primary care with ailments that they can never resolve, such as headache, stomach pain, chest pain, and shortness of breath. Escobar et al. researched and promoted the inclusion of mental health professionals in primary care.

Community Services for Latinos with a Cultural Approach

Reports from government agencies give us information about the use and availability of Hispanic mental health services. From SAMHSA report 2018 [15]:

- 2 in 9 Hispanics have a serious mental illness.
- 43.2% of Hispanic young adults with severe mental illness (SMI) received treatment.

- 60.4% of Hispanic Adults (ages 26–49) with SMI received treatment.

Information about substance use and Hispanics does not differ regarding disparities. For example, among those who needed treatment for drug use and alcohol, Hispanics were less likely than non-Hispanics to receive specialty treatment (data from 2003 to 2011) [16].

Unfortunately, treatment gaps remain vast. Historically Hispanics have not received mental health services as needed. For instance, over the years, there has not been a comprehensive description of Hispanic mental health services. Here are a couple of examples of the first mental health services for Latinos in the literature:

- 1979, "La Frontera" South Tucson, Arizona [17]: a clinic in Arizona that provides mental health services to Mexican Americans.
- 1974, "Hispanic Clinic" New Haven, Connecticut [18]: a pioneering development of mental health services serving Spanish speaking minorities. A collaboration from the state government and an academic institution supports the clinic. Multiple factors increased and improved the services provided at this clinic. The underutilization of services by Hispanics prompted a close relationship with the community to increase engagement, developing the clinic as a trusted place. Also, the development of programs was informed by feedback and continuous communication with religious communities and leaders, local Hispanic newspapers, radio stations in Spanish, and Hispanic related community agencies. The community outreach via conferences and family programs helped to tailor the services to the community needs. The development of links to useful agencies, legal education, substance use education, and political information persists until the present time. When a service expansion was planned, these connections facilitated the formation of a network to create the Connecticut Latino Behavioral Health Services (LBHS), providing bilingual and bicultural clinicians to chosen community agencies for clinical discussions and outcome measuring [19].

Reports about admissions from facilities providing services in Spanish inform about the current delivery of services to Latinos in the United States. A report from 2010 cited that 40.7% of all admissions of Hispanic origin for substance use treatment were in the West, and only 38.5% of all facilities providing services in Spanish were located in that region. The West and North East account for 79.7% of all Hispanic admissions and 65.5% of all facilities that provided behavioral health services in Spanish [20] (Fig. 12.1).

In more recent history, after the worldwide COVID-19 pandemic preventing measures forced providers to use telehealth to follow patient care. Few reports on telehealth considering Hispanics have been published by this time. One study

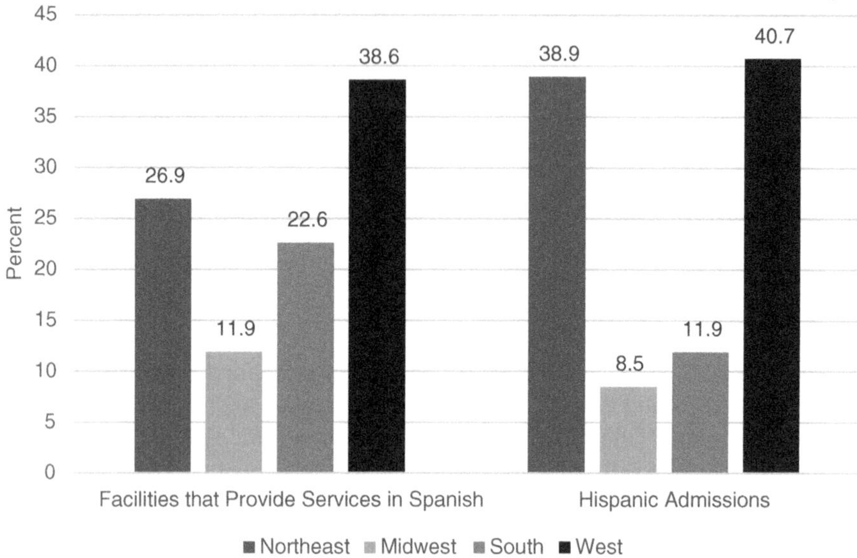

Fig. 12.1 Substance abuse treatment facilities that provide services in Spanish and admission of Hispanic origins, by region. Adapted from SAMHSA 2010

describes a favorable reception, but another pointed out that lack of insurance preclude the use of telehealth services [21].

Influences Promoting the Development of Hispanic Mental Health Services

Research

When developing culturally sensitive services, we have to confront the reality that one of the significant barriers for Hispanics in the United States is language. Besides affecting the quality of services, language is also an obstacle to obtaining information about sources of mental healthcare [22]. According to the 2017 Census, 30% of Hispanics are not proficient in English, and it is noteworthy that age and gender affect this number. Along with studies recommending that the therapist needs to look like the person in treatment [23], health service professionals need to work to address research, mental health services, and education focused on Hispanics [24].

The focus of Latino community mental health interventions turned to underutilization of services and how to increase access to care [25]. It was described that perceptions of mental illness vary culturally; thus, access to mental healthcare varies by ethnic group [22]. For instance, Puerto Ricans have a higher rate of utilization, and in general, Latinos tend to consult mental health problems in primary or family care [14]. Access continues to be a significant problem due to economic constraints, lack of insurance, citizenship limitations, no information about where to go, and staying away from care [26].

Several authors report that Mexican American immigrants in the USA are healthier than Americans and that their offsprings were not as healthy as them, resembling more the American population. In other words, life expectancy is longer for Hispanics from abroad than Americans and second-generation Hispanics. This phenomenon is also known as the Latino Paradox. Moreover, it is suggested that the lack of extended families and social support might be essential influences in this finding. On the other hand, the Latino Paradox does not seem to affect Puerto Ricans, since this group has the worst health of Hispanic groups with increased cardiovascular disorders. A meta-analysis [27] suggests that there is enough evidence of the influence of the Latino Paradox on the health of Hispanics, and there is a need to move beyond and explore risks and resilience mechanisms for these differences. A commentary about the Latino paradox [28] emphasizes the importance of examining the data by country, as the political and socioeconomic variables are different.

The National Latino and Asian American study (NLAAS) is a nationally representative community household survey estimating the prevalence of mental disorders and rates of mental health service utilization by Latinos and Asian Americans. The aim was to assess associations among social position, environmental context, and psychosocial factors with the prevalence of psychiatric disorders and compare to a national representative sample of non-Latino Whites and African Americans. NLAAS was a formal research effort to investigate cultural, ethnic, and environmental considerations often ignored in mental health research. The survey includes the major categories of DSM diagnosis as well as nervous attacks, language preference, discrimination, migration status, and history [29].

The scarcity of Hispanic mental health professionals is a target for intervention. Efforts to increase the racial and ethnic diversity in psychiatry include training of minorities from diverse backgrounds [30], encouraging homegrown candidates by offering support such as scholarships, assure the presence of role models of similar ethnicity, include minorities in boards or administrative positions, advocate for changes, and political involvement [31]. Diversity initiatives address the development of teaching, research, services, and recruitment of minorities [32, 33]. These initiatives aim to assure that educational leadership is diverse and that the new generations of health providers consist of a diverse group representing minorities and underrepresented groups.

Education

Reports from the early 1990s described the scarcity of Latino professionals able to speak Spanish and understand the Latino culture. For example, the Psychology Professional groups reported only 1% of Spanish speaking mental health providers at that time. The scarcity was magnified by the lack of culturally sensitive services for Hispanics; therefore, the efforts focused on cultural competence.

There are many models to teach cultural competence. The Tulane model [34] offers core concepts to address, avoid stereotypes of individuals and groups, learn to ask the right questions, and connect the cultural competence teaching with the clinical skills learned. Kirmayer provides a comprehensive set of educational experiences in cultural psychiatry [35], and by using social science, they aid the trainee to appreciate their backgrounds and to understand culture, ethnicity, and identity. Then the trainee learns to make cultural evaluations as part of all psychiatric assessments [33].

Understanding the ethnic identity and acculturation process that immigrants go through is crucial to provide culturally sensitive psychotherapy and psychiatric treatment [36, 37]. Respect to Hispanic values promotes easy access to treatment and promotes trust. Cultural constructs of the Hispanic population, such as empathy and honesty, are essential for engagement [38, 39]. Support for immigrants should address community and family support, acculturative stress, discrimination, migration history, history of trauma exposure, idioms of distress and resilience, native language and communication preference, and origin [19].

The revision of the DSM IV-TR prompted a look at the outline of the Cultural formulation published in 1994 [12]. In response to requests from clinicians to facilitate cross-cultural evaluations, the Cultural Formulation Interview (CFI) was developed. The use of the Spanish CFI provides information about the unique aspects of the person consulting and his/her culture. By learning the CFI themes and questions, the trainee becomes an expert in eliciting culturally relevant information in all psychiatric assessments. The use of the CFI questions aim that the person consulting reveals the belief of what causes the problem, how do they call it, a description to a friend, their identity, barriers to care, stressors, structural challenges, past and current supports, and their treatment preference [39].

The Inclusion of Culture in Mental Health Services for Latinos

Overall, one of the interventions that increase culturally sensitive services includes using bilingual health providers [40]. Bilingual and bicultural professionals can understand mental health illness from a cultural approach. Studies report that matching patients and providers by the same ethnicity and language increases the rate of resolution of symptoms and increases the patient–provider alliance. Other interventions include adding educational materials in Spanish, embedding cultural values, norms and preferences in the services, and developing tools and training materials to increase staff knowledge and skills regarding mental illness [40, 41].

A successful program that provides culturally oriented and recovery-focused services to the Latino community is the Connecticut Latino Behavioral Health System (LBHS), developed as a culturally informed community-academic collaboration focused on mental health, addiction, and behavioral health [42]. Main points and lessons from the creation and ongoing delivery of services are the inclusion of cultural Latino values, considering recovery the primary goal of treatment, increasing access to services, and promoting the collaboration between community organizations and a state mental health facility [42].

In regard to new programs and the use of technology in mental health, there is no much research in the area of apps, websites, or e-tools in Spanish. Younger generations tend to look towards technology when trying to find more information or referrals to mental health [43, 44]. This lack of services contributes to the gap of care received by the Latino community [45]. Other barriers that contribute to the disparities in access to new programs include the insufficient funding of community programs that would deliver these services to Hispanics [46]. On the other hand, older adults also suffer from disparities in accessing appropriate mental healthcare [38]. Culturally sensitive mental health services focused on recovery must overcome the barriers affecting the older Latino population, such as identifying their particular needs, making services available and accessible, and acceptance of mental healthcare treatment [38, 47].

One of the most important cultural values in the Latino community is *familismo* (familism), the value that refers to warm, close, and supportive family relationships, and that the individual would prioritize the family before themselves [48–50]. Consequently, including and promoting family participation in the treatment of mental illness and the pursue of recovery in mental health is essential [39, 44, 51, 52].

Regarding peer support in mental health services to the Latino community, the lack of Spanish speaking peers is unfortunate and detrimental to the recovery of this community affected by mental health problems. Studies have shown the positive impact of incorporating peer support in the Latino community to new mothers in perinatal care [51]. Utilizing services with same culture peer support increased retention of participants and enhanced participant motivation [53].

Cultural competence in outpatient and inpatient providers is a crucial component of culturally sensitive services and their impact on patients' recovery [52, 54]. For example, the participation of culturally competent health providers in health programs has shown to increase engagement in treatment, retention, and response in family-based interventions [55, 56].

Benefits of Implementing and Protecting Cultural Sensitive Services for Latinos

The two most significant benefits of implementing culturally sensitive services for Latinos relate to decreasing barriers to engage and complete treatment [57] as well as increasing the therapeutic alliance between provider and patient [58].

Culturally sensitive services aim at the engagement and retention of patients in treatment. Latinos tend to be skeptical of diagnostic labels and the benefits of standard forms of mental health treatment [59]. Successful culturally sensitive programs target the gap between ethnic and medical culture as it has been shown that the differences between provider–patient explanatory models of illness are a barrier in accessing mental healthcare [60].

Independence and the belief of being able to treat mental health issues with home remedies is another vital factor related to Latino immigrants and access to treatment [61]. Programs that address Latino's self-sufficient attitudes of behavioral health, a cultural aspect of the Latino community, may increase initiation and reduce attrition from treatment [62]. Concerning ways of expressing distress and understanding of symptoms, Latinos are more likely to conceptualize mental illness as a somatic problem [59]. Therefore, the models that emphasize behavioral change interventions that target physical ailments may seem less stigmatizing and more culturally relevant because they match Latinos' somatic beliefs about mental illness [63].

Culturally sensitive programs take advantage of the strong social ties in the Latino community to engage patients with treatment. Though social collectivism can make engagement in treatment among Latinos a more laborious process [64], peer navigator programs take advantage of this cultural element to enhance service engagement and recovery among Latinos with serious mental illness [65].

Successful culturally sensitive programs breach the gap between language and culture. Though language is a barrier to access care among Latino immigrants [61], mastery of Spanish without an understanding and curiosity for specific cultural details might lead to less successful treatment results. For example, culture-bound syndromes such as *mal de ojo* or *ataque de nervios* can be better understood when a provider is culturally aware of the significance of those terms in the Latino community [66]. Latino proverbs and sayings (known as *dichos*) are another essential cultural element that culturally competent providers can use to connect with patients. As outlined by psychologist Marlene de Rios, *dichos* "can link the phenomenological world of the Latino immigrant with the symbols and metaphors available" [67], which in turn can create a stronger therapeutic alliance between provider and patient.

Challenges Encountered When Incorporating Culturally Sensitive Services

Despite the many benefits that culturally sensitive services can provide, unfortunately, these competences are not widespread across the country. One of the main challenges in implementing these services is the availability of qualified staff to provide local training and implementing these programs. Highly successful, culturally sensitive programs rely on bilingual staff to adapt and adequately tailor successful behavioral models into culturally sensitive interventions. Unfortunately, the number of bilingual mental health providers in the United States is not enough to widely disseminate these programs [68]. Without adequate qualified bilingual trainers, it is difficult to emphasize and disseminate the importance of not only language but culture-bound nuisances of mental health seen in the Latino community.

Adapting culturally sensitive services to the Hispanic community has its challenges; however, it is essential. Maintaining content equivalency between the original intervention and the culturally sensitive intervention can prove challenging and raise into question the validity of the latter [69]. This difficulty can be minimized by using a systematic approach that documents the different adaptations and focuses on understanding theoretical driven principles and processes instead of making a rigid translation of the first intervention [70]. By following the above, cultural adaptation models of services can provide flexible guidelines to work with minority populations, such as Latinos [70].

In the presented context, another barrier to the delivery of culturally sensitive services is the diversity within the Latino community. Even though most Central and South America share the same language, specific culture-bound mental health syndromes, for example, "*susto,*" appear to have different treatments and descriptions among different regions [71]. The above leads for a need to further specialize in different interventions based on specific Latino demographics, which, in turn, makes it more challenging to implement these programs on a national scale widely.

Lastly, implementing any kind of additional training incurs financial and time constraints. Persuading administrators of an already financially burden medical system requires cultural humility that might not always be present. It is presumed that as cultural sensitivity becomes more widespread and accepted as a norm, the implementation of culturally sensitive services can be more warmly welcomed.

Advocacy for Development and Maintenance of Culturally Sensitive Services for Hispanics

The development of services has been guided by advocacy from multiple disciplines to address policies, education, research, and funding for Hispanic mental health services [Table 12.1].

Professional groups providing mental health services have debated how to assure that all people from all races and Culturally And Linguistically Diverse (CALD) populations receive equitable and effective healthcare. Formal efforts to address cultural sensitivity to minorities started in 1980; it was named cultural competence. The guidelines from the Accreditation Council for Graduate Medical Education (ACGME) competencies related to cultural competence, as well as the American Academy of Child and Adolescent Psychiatry's (AACAP) recommendations for the cultural competence training of child/adolescent fellows are useful. However, they have not solved the problem of health disparities and the scarcity of minorities in Psychiatry.

Table 12.1 Funding sources for Hispanic mental Health

Affordable Care Act (ACA)	After the ACA Hispanics, had the most significant percentage increased in health coverage from 32.6% to 19.1% in the states where the expansion took place [26, 72]. This coverage affected Hispanics eligible for Medicaid. Undocumented Immigrants do not qualify for government coverages and continue to struggle with disparities in healthcare.
Substance Abuse and Mental Health Services Administration (SAMHSA)	SAMHSA Grants provide financial support for agencies to increase substance use and behavioral health services to Latinx yearly. They also provide education on evidence-based medicine and culturally competent services. SAMHSA website provides a behavioral health treatment services locator, but they do not identify what agencies offer treatment in Spanish. The seeker would have to call the agency to obtain that information. https://findtreatment.samhsa.gov
National Alliance on Mental Illness (NAMI)	NAMI provides support and information in Spanish for people in need of mental healthcare and their families. https://www.nami.org/Your-Journey/Identity-and-Cultural-Dimensions/Latinx-Hispanic/La-salud-mental-en-la-comunidad-latina
National Institute of Mental Health (NIMH)	NIMH provides research support to understand healthcare by ethnicity, study health outcomes, and try new interventions.
American Psychiatric Association (APA)	The APA, specifically the Hispanic caucus, aims to address mental health services for Hispanics and to educate the group
American Society of Hispanic Psychiatry (ASHP)	ASHP is a professional organization of mental health professionals interested in supporting the development of Hispanic services and research since 1982. This group was created to address research, services, and education in Latino mental health professionals. Senior members provide guidance and mentorship to younger generations to advance the field of Hispanic psychiatry and contribute to the pipeline of bilingual, bicultural, psychiatrists. The latter could provide not only adequate services but also do research.

Spurlock leadership was the first formal intervention to promote a diverse workforce. The minority fellowship program was funded in 1974 by the National Institute of Mental Health to train underrepresented minorities, contributing to diverse and culturally sensitive professionals. Examples of successful outcomes are the formation of scholarly networks on Hispanic mental health [73] and the support to create new educational programs in medical schools of states in need of culturally sensitive psychiatrists.

Conclusions

Delivery of culturally sensitive mental health services helps provide high quality and appropriate care. Most care models are developed in a non-Hispanic culture; therefore, these services need to be translated and culturally adapted to the population that will serve. The bilingual and bicultural professionals should take the task of adapting and translating such models.

As mentioned in this chapter, continued research, and clinical and community programs focused on the Hispanic culture are a crucial piece to develop services that help equally to this population and consider culture as a principal pillar [74]. However, in order to provide the best care to Hispanic patients, culturally sensitive mental health services must be accompanied by social justice, advocacy, dissemination of research, and policies [75]. As mental health professionals, we must develop and protect those tools in order to fulfill the essential needs of the Latino population.

References

1. Alegría M, Canino G, Ríos R, Vera M, Calderón J, Rusch D, et al. Inequalities in use of specialty mental health services among Latinos, African Americans, and non-Latino whites. Psychiatr Serv. 2002;53(12):1547–55.
2. Burnett-Zeigler I, Lee Y, Bohnert KM. Ethnic identity, acculturation, and 12-month psychiatric service utilization among black and Hispanic adults in the U.S. J Behav Health Serv Res. 2018;45(1):13–30.
3. López SR, Barrio C, Kopelowicz A, Vega WA. From documenting to eliminating disparities in mental health care for Latinos. Am Psychol. 2012;67(7):511–23.
4. Chang C-W, Biegel DE. Factors affecting mental health service utilization among Latino Americans with mental health issues. J Ment Health. 2018;27(6):552–9.
5. Margarita Alegría PD, Glorisa Canino PD, Ruth Ríos PD, Mildred Vera PD, José Calderón MS, Dana Rusch BA, et al. Mental health care for Latinos: Inequalities in use of specialty mental health services among latinos, African Americans, and non-Latino whites. Psychiatr Serv. 2002;53(12):1547–55.
6. Ortega AN, McKenna RM, Kemmick Pintor J, Langellier BA, Roby DH, Pourat N, et al. Health care access and physical and behavioral health among undocumented Latinos in California. Med Care. 2018;56(11):919–26.
7. McGinty EE. Interventions to address medical conditions and health-risk behaviors among persons with serious mental illness: a comprehensive review, Schizophrenia Bulletin. 2016;42(1)
8. Kirmayer LJ, Young A. Culture and somatization: clinical, epidemiological, and ethnographic perspectives. Psychosom Med. 1998;60(4):420–30.
9. Alarcon RD, Bell CC, Kirmayer LJ, Lin K-M, Ustun B, Wisner KL. In: Kupfer DJ, editor. Beyond the funhouse mirrors: Research agenda on culture and psychiatric diagnosis; 2002. p. 219–81.
10. Mezzich JE, Ruiz P, Munoz RA. Mental health care for Hispanic Americans: a current perspective; 1999. p. 91–102.
11. Kirmayer LJ. The place of culture in psychiatric nosology: Taijin kyofusho and DSM-III-R. J Nerv Ment Dis. 1991;179(1):19–28.
12. Mezzich JE, Kirmayer LJ, Kleinman A, Fabrega H Jr, Parron DL, Good BJ, et al. The place of culture in DSM-IV. J Nerv Ment Dis. 1999;187(8):457–64.
13. Lewis-Fernández R, Aggarwal N, Lam P, Galfalvy H, Weiss M, Kirmayer L, et al. Feasibility, acceptability and clinical utility of the cultural formulation interview: mixed-methods results from the DSM-5 international field trial. Br J Psychiatry. 2017;210(4):290–7.
14. Escobar JI, Rubio-Stipec M, Canino G, Karno M. Somatic symptom index (SSI): a new and abridged somatization construct. Prevalence and epidemiological correlates in two large community samples. J Nerv Ment Dis. 1989;177(3):140–6.

15. Administration SAaMHS. National Survey on drug use and health: Hispanics, Latino or Spanish Origin or Descent. Rockville MD U.S. Department of Health and Human Service, 2018.
16. Substance Abuse and Mental Health Services Administration. Results from the 2018 National Survey on Drugs Use and Health: Detailed Tables. 2019 (Center for Behavioral Health Statistics and Quality, Rockville, Maryland 20857).
17. Martin PP. La Frontera perspective: Providing mental health services to Mexican Americans Tucson Az1979.
18. Abad V, Ramos J, Boyce E. A model for delivery of mental health services to Spanish-speaking minorities. American Journal of Orthopsychiatry. 1974;44:584–95.
19. Paris M, Silva MA, Diaz E, Bedregal LE, Cole RA, Añez-Nava L. The Connecticut Latino behavioral health system: a culturally informed community-academic collaboration. Psychol Serv. 2016;13(2):140–7.
20. Quality CfBHSa. Substance Abuse Facilties with Services in Spanish are available in all regions for Hispanic clients admitted to treatment in 2010. Human Health Services: Rockville MD; 2010.
21. Hilty D, Lim R, Nasatir-Hilty S, Koike A, Ton H, Nesbitt T. Planning for Telepsychiatric consultation: a m=needs assessment for cultural and language Services at Rural Sites in California. J Rural Ment Health. 2015;39(3–4):153–61.
22. Martinez Tyson D, Arriola N, Corvin J. Perceptions of depression and access to mental health care among Latino immigrants: looking beyond one size fits all. Qualitative Health Reserach. 2016;26(9):1289–302.
23. Sue S. In search of cultural competence in psychotherapy and counseling. Am Psychol. 1998;53(4):440–8.
24. Services USDoHaH. Mental Health: Culture, Race, and Ethnicity–A Suppl to Mental Health: A Report of the Surgeon General. Rockville, Md.: 2001.
25. Vega WA, Kolody B, Aguilar-Gaxiola S, Catalano R. Gaps in service utilization by Mexican Americans with mental health problems. Am J Psychiatry. 1999;156(6):928–34.
26. Jennifer T, Kendal O, Natalie S, Anthony D. Key facts about the uninsured population. DC: Washington; 2019.
27. Ruiz JM, Steffen P, Smith TB. Hispanic mortality paradox: a systematic review and meta-analysis of the longitudinal literature. Am J Public Health. 2013;103(3):e52–60.
28. Abraído-Lanza AF, Dohrenwend BP, Ng-Mak DS, Turner JB. The Latino mortality paradox: a test of the "salmon bias" and healthy migrant hypotheses. Am J Public Health. 1999;89(10):1543–8.
29. Alegria M, Takeuchi D, Canino G, Duan N, Shrout P, Meng X-L, Vega W, Zane N, Vila D, Woo M, Vera M, Guarnaccia P, Aguilar-Gaxiola S, Sue S, Escobar J, Lin K-M, Gong F. Considering context, place and culture: the National Latino and Asian American study. Int J Methods Psychiatr Res. 2004;13(4):208–20.
30. Lokko N, Chen J, Parekh R, Stern T. Racial and ethnic diversity in the US psychiatric force: a perspective and recommendations academic psychiatry 2016.
31. Statement on the Better Care Reconciliation, (2017).
32. Lim R, Luo J, Suo S, Hales R. Diversity initiatives in academic psychiatry: applying cultural competence. Acad Psychiatry. 2008;32:283–90.
33. Lim RF, Lu FG. Culture and psychiatric education. Acad Psychiatry. 2008;32(4):269–71.
34. Pena JM, Manguno-Mire G, Kinzie E, Johnson JE. Teaching cultural competence to psychiatry residents: seven Core concepts and their implications for therapeutic technique. Acad Psychiatry. 2016;40(2):328–36.
35. Kirmayer LJ, Groleau D, Guzder J, Blake C, Jarvis E. Cultural consultation: a model of mental health Service for Multicultural Societies. The Canadian Journal of Psychiatry/La Revue canadienne de psychiatrie. 2003;48(3):145–53.
36. Delgado-Romero EA, Hernandez CA. Empowering Hispanic students through student organizations: competencies for faculty advisors. J Hisp High Educ. 2002;1(2):144–57.

37. Akhtar S. Immigration and acculturation mourning, adaptation, and the next generation. Lanham: Jason Aronson; 2011. p. xvi, 294.
38. Costantino G, Malgady RG, Primavera LH. Congruence between culturally competent treatment and cultural needs of older Latinos. J Consult Clin Psychol. 2009;77(5):941–9.
39. Diaz E, Anez LM, Silva M, Paris M, Davidson L. Using the cultural formulation interview to build culturally sensitive services. Psychiatr Serv. 2017;68(2):112–4.
40. Lewis-Fernández R, Aggarwal NK, Bäärnhielm S, Rohlof H, Kirmayer LJ, Weiss MG, et al. Culture and psychiatric evaluation: operationalizing cultural formulation for DSM-5. Psychiatry. 2014;77(2):130–54.
41. Cabassa LJ, Manrique Y, Meyreles Q, Capitelli L, Younge R, Dragatsi D, et al. "Treated me ... like I was family": qualitative evaluation of a culturally-adapted health care manager intervention for Latinos with serious mental illness and at risk for cardiovascular disease. Transcult Psychiatry. 2019;56(6):1218–36.
42. Paris M, Silva MA, Diaz E, Bedregal LE, Cole RA, Anez-Nava LM. The Connecticut Latino behavioral health system: a culturally informed community-academic collaboration. Psychol Serv. 2016;13(2):140–7.
43. Chaet AV, Morshedi B, Wells KJ, Barnes LE, Valdez R. Spanish-language consumer health information technology interventions: a systematic review. J Med Internet Res. 2016;18(8):e214.
44. Kapke TL, Gerdes AC. Latino family participation in youth mental health services: treatment retention, engagement, and response. Clin Child Fam Psychol Rev. 2016;19(4):329–51.
45. Ospina-Pinillos L, Davenport T, Mendoza Diaz A, Navarro-Mancilla A, Scott EM, Hickie IB. Using Participatory Design Methodologies to Co-Design and Culturally Adapt the Spanish Version of the Mental Health eClinic: Qualitative Study. J Med Internet Res. 2019;21(8):e14127-e.
46. Rosales R, Calvo R. Characteristics of Hispanic-serving mental health organizations in the U.S. J Health Care Poor Underserved. 2019;30(2):547–59.
47. Guzman ED, Woods-Giscombe CL, Beeber LS. Barriers and facilitators of Hispanic older adult mental health service utilization in the USA. Issues Ment Health Nurs. 2015;36(1):11–20.
48. Campos B, Ullman JB, Aguilera A, Dunkel SC. Familism and psychological health: the intervening role of closeness and social support. Cultur Divers Ethnic Minor Psychol. 2014;20(2):191–201.
49. Ayón C, Aisenberg E. Negotiating cultural values and expectations within the public child welfare system: a look at familismo and personalismo. Child Fam Soc Work. 2010;15(3): 335–44.
50. Calzada EJ, Tamis-LeMonda CS, Yoshikawa H. Familismo in Mexican and Dominican families from low-income, urban communities. Journal of Family Issues. 2012;34(12): 1696–724.
51. Valdez CR, Ramirez Stege A, Martinez E, D'Costa S, Chavez T. A community-responsive adaptation to reach and engage Latino families affected by maternal depression. Fam Process. 2018;57(2):539–56.
52. Villatoro AP, Morales ES, Mays VM. Family culture in mental health help-seeking and utilization in a nationally representative sample of Latinos in the United States: the NLAAS. Am J Orthopsychiatry. 2014;84(4):353–63.
53. Lutenbacher M, Elkins T, Dietrich MS, Riggs A. The efficacy of using peer mentors to improve maternal and infant health outcomes in Hispanic families: findings from a randomized clinical trial. Matern Child Health J. 2018;22(Suppl 1):92–104.
54. Sanchez K, Killian MO, Eghaneyan BH, Cabassa LJ, Trivedi MH. Culturally adapted depression education and engagement in treatment among Hispanics in primary care: outcomes from a pilot feasibility study. BMC Fam Pract. 2019;20(1):140.
55. Kumpfer KL, Alvarado R, Smith P, Bellamy N. Cultural sensitivity and adaptation in family-based prevention interventions. Prev Sci. 2002;3(3):241–6.

56. Semansky RM, Altschul D, Sommerfeld D, Hough R, Willging CE. Capacity for delivering culturally competent mental health services in New Mexico: results of a statewide agency survey. Admin Pol Ment Health. 2009;36(5):289–307.
57. Betancourt JR, Green AR, Carrillo JE, Ananeh-Firempong O 2nd. Defining cultural competence: a practical framework for addressing racial/ethnic disparities in health and health care. Public Health Rep. 2003;118(4):293–302.
58. Brach C, Fraser I. Can cultural competency reduce racial and ethnic health disparities? A review and conceptual model. Med Care Res Rev. 2000;57(Suppl 1):181–217.
59. Carpenter-Song E, Chu E, Drake RE, Ritsema M, Smith B, Alverson H. Ethno-cultural variations in the experience and meaning of mental illness and treatment: implications for access and utilization. Transcult Psychiatry. 2010;47(2):224–51.
60. Cortes DE, Mulvaney-Day N, Fortuna L, Reinfeld S, Alegría M. Patient--provider communication: understanding the role of patient activation for Latinos in mental health treatment. Health Educ Behav. 2009;36(1):138–54.
61. Falgas I, Ramos Z, Herrera L, Qureshi A, Chavez L, Bonal C, et al. Barriers to and correlates of retention in behavioral health treatment among Latinos in 2 different host countries: the United States and Spain. J Public Health Manag Pract. 2017;23(1):e20–e7.
62. Ransford HE, Carrillo FR, Rivera Y. Health care-seeking among Latino immigrants: blocked access, use of traditional medicine, and the role of religion. J Health Care Poor Underserved. 2010;21(3):862–78.
63. Jimenez DE, Burrows K, Aschbrenner K, Barre LK, Pratt SI, Alegría M, et al. Health behavior change benefits: perspectives of Latinos with serious mental illness. Transcult Psychiatry. 2016;53(3):313–29.
64. Hough RL, Hazen AL, Soriano FI, Wood P, McCabe K, Yeh M. Mental health services for Latino adolescents with psychiatric disorders. Psychiatr Serv. 2002;53(12):1556–62.
65. Corrigan P, Sheehan L, Morris S, Larson JE, Torres A, Lara JL, et al. The impact of a peer navigator program in addressing the health needs of Latinos with serious mental illness. Psychiatr Serv. 2018;69(4):456–61.
66. Romero AJ, Edwards LM, Corkery S. Chapter 17–assessing and treating Latinos: overview of mental Health Research. In: Paniagua FA, Yamada A-M, editors. Handbook of multicultural mental health. 2nd ed. San Diego: Academic Press; 2013. p. 327–43.
67. de Rios MD. Brief psychotherapy with the Latino immigrant client. New York, NY, US: Haworth Press; 2001. p. 191.
68. Demographics of U.S. Psychology Workforce [Interactive Data Tool] [Internet]. 2019 [cited 07/05/2020]. Available from: https://www.apa.org/workforce/data-tools/demographics
69. Bernal G, Bonilla J, Bellido C. Ecological validity and cultural sensitivity for outcome research: issues for the cultural adaptation and development of psychosocial treatments with Hispanics. J Abnorm Child Psychol. 1995;23(1):67–82.
70. Domenech Rodríguez MM, Baumann AA, Schwartz AL. Cultural adaptation of an evidence based intervention: from theory to practice in a Latino/a community context. Am J Community Psychol. 2011;47(1–2):170–86.
71. Weller SC, Baer RD, de Alba Garcia JG, Glazer M, Trotter R, Salcedo Rocha AL, et al. Variation and persistence in Latin American beliefs about evil eye. Cross-Cult Res. 2014;49(2):174–203.
72. Artiga S, Orgera K, Damico A. Changes in Health care coverage by Race and Ethnicity since the ACA, 2010-2018. KFF, 2020.
73. Harris B, Mian A, Lomax J, Scott-Gurnell K, Sargent J, Phillips J, et al. The Texas regional psychiatry minority Mentor network: a regional effort to increase Psychiatry's workforce diversity. Acad Psychiatry. 2012;36(1):60–3.
74. Manuel JI. Racial/ethnic and gender disparities in health care use and access. Health Serv Res. 2018;53(3):1407–29.
75. Torres SA, Santiago CD, Walts KK, Richards MH. Immigration policy, practices, and procedures: the impact on the mental health of Mexican and central American youth and families. Am Psychol. 2018;73(7):843–54.

Medical Comorbidities of Hispanics with Mental Health Problems

13

David Martinez Garza, Elliott R. Weinstein, and Daniel E. Jimenez

Individuals with chronic physical health conditions are more likely to have a mental health disorder than those without chronic physical health conditions [1–3]. The co-occurrence of mental and physical illness (i.e., comorbidity) incrementally worsens health compared with mental illness alone, chronic disease alone, and with any combination of chronic diseases without mental illness [4]. Previous studies suggest that when a mental illness is comorbid with a physical illness, the comorbidity is associated with greater disease severity, high degree of functional impairment, increased health care services needed to manage the comorbid conditions; decreased quality of life, and mortality [5–9].

Racial/ethnic differences in the full array of comorbid conditions are not available, but we do know that the physical and mental health of Hispanics is complicated by age, subethnicity (i.e., Mexican, Cuban, Puerto Rican, etc.), and time in the United States. National surveys that combine Hispanic adults into one large, homogeneous group indicate that Hispanics are at lower risk for psychiatric disorders such as major depressive disorder and generalized anxiety disorder compared to non-Hispanic whites [10–12]. However, if taken at face value, this might convey the false impression that Hispanics are healthier (i.e., immigrant paradox) rather than the more nuanced picture when subethnicity, age, and time in the United States are contextualized.

D. M. Garza
Department of Psychiatry, Massachuessetts General Hospital, Boston, MA, USA

E. R. Weinstein
Department of Psychology, University of Miami, Florida, USA

D. E. Jimenez (✉)
Department of Psychiatry and Behavioral Sciences, University of Miami Miller School of Medicine, Florida, USA
e-mail: dej18@miami.edu

R. Castilla-Puentes, T. Falcone (eds.), *Mental Health for Hispanic Communities*,
https://doi.org/10.1007/978-3-031-13195-0_13

Some ethnic subgroups have higher rates of psychiatric diagnoses, such as Puerto Ricans and Cubans. In a study comparing lifetime and 12-month prevalence of the fourth edition of the Diagnostic and Statistical Manual (DSM-IV) psychiatric disorders, Hispanics and non-Hispanic whites had similar lifetime prevalence rates of any depressive disorder (16.4% vs. 12.2%) and of any anxiety disorder (15.3% vs. 13.5%; [13]). When stratified by time in the United States, recent immigrants had the lowest rates of mental health comorbidities, yet the risk increases in a time-dependent manner with longer stays, reaching a very close approximate to the US general population after 13 years of living in the United States. The same is true for medical comorbidities, as recent migrant Hispanics, as a whole, have a high risk for certain chronic medical conditions but low for other conditions. For example, Hispanics have a greater prevalence of diabetes and obesity but lower prevalence of cardiovascular disease and cancer compared to non-Hispanic whites [14–19]. This difference tends to disappear over time. The results of these two studies suggest that immigrant health advantages earlier in life yield to the cumulative adversity of homeland estrangement, social isolation, and the overwhelming effects of socioeconomic disadvantages in later years when health needs and costs rise sharply [20]. This is most likely secondary to a breakdown of the family network and social support systems the longer migrants reside in the United States, as well as to acculturation and adaptation of American lifestyles (e.g., diets high in fat, salt, and cholesterol, smoking patterns, and sedentarism) and traditions (e.g., less of an emphasis in familism, and different family bonds and social values).

As clinicians, it is critical to understand that Hispanics and their subgroups may show very different patterns of psychiatric and medical comorbidities, as well as different responses to preventive measures, treatment, and prognosis relative to non-Hispanic white or Black populations. The purpose of this chapter is two-fold. First, we provide the reader with a more in-depth picture of the role that medical comorbidities play among the different Hispanic subgroups with mental health disorders. To describe the full array of comorbid conditions would be beyond the scope of this chapter; therefore, we focus on six comorbid conditions that consistently impact the mental health of Hispanics. Second, we provide treatment recommendations from a psychiatric and psychological perspective on how to treat Hispanics given these comorbidities. Throughout the chapter, we make clear when the research is based on Hispanics as a whole or on specific subgroups.

Cardiovascular Disease

Cardiovascular disease (CVD) is the leading cause of death in the United States. One in three adults over the age of 35 years ultimately dies of cardiovascular disease. These include, but are not limited to, hypertension, atherosclerosis, coronary artery disease (CAD), myocardial infarction (MI), conduction abnormalities, peripheral artery disease, and the most common final pathway of the above, congestive heart failure (CHF). The symptoms of these can vary from none (as in the case of hypertension) to chest pain, fatigue, peripheral edema, and to debilitating shortness of breath at rest in end-stage CHF. Hispanics have a higher incidence of heart failure compared with non-Hispanic whites, and Hispanics who present with heart failure are younger with more comorbidities [21–23]. However, Hispanics represent a relatively young demographic such that the future public health burden of clinical heart failure as this population ages is potentially underestimated.

Psychiatric patients are specifically vulnerable to CVD given a number of risk factors. Behavioral patterns associated with mental health conditions such as smoking, a sedentary lifestyle, poor nutrition, and a lack of compliance with treatment increase the risk of developing CVD. Most antipsychotics and many antidepressants carry unfavorable cardiometabolic side effect profiles such as weight gain, and increased risk of developing laboratory abnormalities such as elevated glucose and lipids. Furthermore, psychological factors such as acute stress and anxiety, depression, and personality profiles of anger and hostility have all been linked to an elevated risk of developing CVD [24].

CVD has a prevalence of 5–10% with co-occurring anxiety disorder (primarily panic disorder) and 10–15% with mood disorders (primarily major depressive and persistent depressive disorders). Moreover, anxiety disorders in general have been found to be predictors of the incidence of hypertension (relative risk of 1.5–3.0) and major depressive disorder occurs in 15–20% after an MI and in up to 40% individuals 6 months after coronary artery bypass surgery. Finally, 15% of patients post MI or post cardiac surgery develop post-traumatic stress disorder (PTSD), which is in and of itself associated with a relative risk of 2.0 of recurrent cardiac events and death. Therefore, patients who suffer an MI and develop PTSD are 2 times more likely to have another MI and die than their peers who do not develop PTSD [24].

Of all psychiatric disorder, depression has been the most studied regarding its link to CVD. Depression is an independent risk factor associated with developing CAD and individuals with depression are 1.5 to 2 times more likely to have an MI and cardiac-related mortality. There appears to be a dose–response relationship between the severity of depression and cardiovascular outcomes. For example, in a prospective study (4.5 years) of more than 4500 patients with CVD risk factors, CHF developed in 8.1% with high depressive symptoms compared to only 3.2% in patients with low depressive symptoms [25]. In another retrospective study, the mortality rate after 5 years was 36% in depressed patients with CHF compared to 16% in non-depressed patients with CHF. These effects appear to persist chronically, according to 10–20-year follow-up longitudinal studies [26]. Depression has also been independently associated with the risk of developing conduction abnormalities, including atrial fibrillation and ventricular arrhythmias with secondary sudden cardiac death [24].

Even though not as thoroughly reviewed as depression, other psychiatric disorders and more recently, PTSD, have also been recognized as both risk factors for and negative outcomes of CVD. Most theories studying this link support a biopsychosocial model in which inherent biological properties of psychiatric illness (such as inflammation and metabolic abnormalities) as well as psychosocial factors such as unhealthy lifestyle habits, poor self-care, and lack of insight and adherence to treatment exacerbate the link between depression and other psychiatric disorders. According to the Updated 2013 Statistical Fact Sheet, published by the American Heart Association, the overall prevalence of any type of CVD among Hispanics over

Table 13.1 Incidence of hypertension stratified by Hispanic subgroup

	Overall	Dominicans	Cubans	Puerto Ricans	Central Americans	South Americans	Mexicans
Men	21.7%	28.1%	27.1%	21.9%	19.8%	18.3%	17.6%
Women	19.7%	23.3%	22.6%	28.2%	18.7%	16.1%	16.0%

20 years old is 33.4% for men and 30.7% for women, with the prevalence of CAD being 6.7% and 5.3%, respectively. In the Hispanic Community Health Study/Study of Latinos (HCHS/SOL), which included data from over 16,000 Hispanics living in four urban US cities, the 6-year incidence of hypertension (measured as a SBP >130 or a DBP >80) was around 20%. Interestingly, but not surprisingly, when you stratify Hispanics into ethnic subgroups, the incidence varies significantly among them (Table 13.1).

While the prevalence of hypertension among Hispanics is lower than their white counterparts (25.5% HCHS/SOL vs 27.4% NHANES), the percentage of Hispanic patients being in any type of pharmacological treatments is actually subpar (63.4% vs 76.6%) with the percentage of patients being adequately controlled dropping even further (37.5% vs 56.3%). These observed disparities highlight the differences in access and quality in which Hispanics are usually at a disadvantage [27–29].

The treatment of psychiatric disorders is generally complicated even more given the aforementioned cardiometabolic risks of many psychotropic agents. Nonetheless, given the poor outcomes associated with co-occurring psychiatric illness, it is critical to identify and aggressively treat these patients. Again, most of the literature available has focused on the treatment of depression in patients with comorbid CVD. Many psychotropics are to be used with caution when patients suffer from cardiac impairment. Of special consideration are agents that have an increased risk of conduction abnormalities, such as most of the tricyclic antidepressants and the selective serotonin inhibitor citalopram at doses higher than 40 mg daily. Venlafaxine, a selective norepinephrine-serotonin inhibitor, and bupropion, a dual dopamine-norepinephrine inhibitor, have both been associated with risk of increasing blood pressure at high doses (>225 mg daily for venlafaxine and > 300 mg for bupropion). In the Sertraline Treatment of Major Depression in Patient with Acute MI or Unstable Angina (SADHART) trial, sertraline demonstrated safety and efficacy in treating depressive symptoms in patients following cardiac events, with a trend towards fewer adverse cardiac events [30]. Recently, the Escitalopram for Depression in Acute Coronary Syndrome (EsDEPACS) study also found escitalopram to be safe and effective in treating depression in patients following an acute coronary syndrome with improved long-term cardiac outcomes [31]. Randomized clinical trials have not been conducted for the treatment of anxiety disorders or PTSD in patients with CVD. Therefore, following general guidelines would be a reasonable approach.

Cerebrovascular Disease

Stroke is a major cause of morbidity and mortality with strokes serving as the second leading cause of death worldwide and the fourth leading cause of death in the United States (5% of deaths are related to stroke). Most strokes (85–90%) are due to atherosclerotic disease with similar risk factor to those discussed in the cardiovascular disease section. Similar to CVD, depression increases the likelihood of having a stroke [32, 33]. The incidence of first-time stroke increases with age with an annual incidence of 2.5–5.3 per 1000 individuals between the ages of 55–64 and increases to 8.9–13.9 per 1000 in the group age of 75–84. The incidence of recurrence stroke also varies with age and is about 7% per year [32]. Hispanics have an overall lower stroke risk than non-Hispanic whites (odds ratio of 0.58; [34, 35]). However, Hispanics, specifically Mexican Americans, have higher rates of transient ischemic attacks and according to a recent survey in a cohort in Corpus Christi, Texas, there is a rise in the trend of incident stroke among Mexican Americans vs non-Hispanic whites (168 vs 136 per 10,000 persons). Given this trend, it is believed that the prevalence of stroke among Hispanics will be equal to that of their non-Hispanic white counterparts in the near future [36, 37].

By far, the most relevant psychiatric comorbidity associated with cerebrovascular disease is post-stroke depression. Most meta-analyses place the incidence of post-stroke depression in the range of 29–31% within the first 5 years after stroke, with some cohorts having as high as 50% of prevalence within the first year; numbers that are substantially higher than the general prevalence of depression among non-stroke patients (7.1%; [33, 38]). Post-stroke depression has a negative effect on functional outcome and is independent of the severity of stroke itself, and more importantly, post-stroke depression is significantly and clinically associated with an increased short- and long-term mortality risk. People with post-stroke depression have a 50% increase in mortality at 1 year and are 3.4 times more likely of dying at 10 years when compared with non-depressed post-stroke patients [38].

The risk of developing other psychiatric comorbidities in addition to depression is also elevated post stroke. For instance, the risk of developing anxiety is 18% (most commonly phobic disorders and generalized anxiety disorder), and the risk of developing both, comorbid anxiety and depression, is 14%. Apathy, the loss of motivation and/or drive, develops in more than one out of every three patients post stroke (34.5%), and it is a characteristic manifestation of recovering stroke patients that is independent of depression [38].

With such high rates, it is clear that strategies to diagnose and treat post-stroke depression early are paramount. The first randomized clinical trials with antidepressant therapy were conducted decades ago, with a superior survival rate at 9 years of 67.9% receiving either fluoxetine or nortriptyline vs 35.7% in the placebo group. Since then, many have been conducted showing similar efficacy results. A Cochrane review by Hackett et al. [39] favored antidepressant therapy over control for post-stroke depression. They found that antidepressants not only increased life expectancy but also had a positive impact on quality of life, with improved motor recovery and decreased rates of disability. The American Heart Association recommends the use of antidepressants for post-stroke depression for at least 6 months beyond recovery, albeit in clinical practice, the maintenance antidepressant therapy is usually used indefinitely due to the high risk of relapse.

Given the high risk of incidence of post-stroke depression, recent clinical trials have explored the use of antidepressant therapy prophylactically. The first such trial evaluated escitalopram vs problem solving therapy (PST) vs placebo in patients following an acute stroke found that after one year, the incidence of post-stroke depression was 8.5% for the placebo group, vs 11.9% and 22.4% for the PST and placebo arms, respectively [40]. Since then, similar trials with matching positive results have been conducted with fluoxetine, sertraline, and milnacipran, with no significant differences in side effects [39, 41]. Even though not a mainstream of treatment, prophylactic antidepressant therapy should be strongly considered for patients with a high number of risk factors for developing post-stroke depression, especially those with a history of mood disorders or those with greater burden of lesion, and if indicated, its preventive use would be reasonable and justified.

Cardiometabolic Diseases

Hispanics are at a disproportionately higher risk of developing cardiometabolic diseases, such as metabolic syndrome (MetS) and diabetes, as well as obesity relative to any other racial/ethnic group in the United States [42, 43]. Moreover, Hispanics adults are more likely to be sedentary, and they are not as actively engaged in pursuing changes in their physical activity compared to non-Hispanic whites [44, 45].

According to the World Health Organization [46], criteria for MetS include insulin resistance and any two of the following: (1) raised arterial pressure, (2) raised plasma triglycerides, (3) low High Density Lipoprotein cholesterol (HDL-C), (4) central obesity or elevated body mass index (BMI), and (5) high urinary albumin excretion rate. MetS emerges in midlife and is related to the development of diabetes, heart disease, and stroke [47]. MetS affects about one-third of United States adults and occurs earlier and more commonly among Hispanics compared with other racial/ethnic groups [48]. In the population-based cohort (HCHS/SOL), 35% of participants met the criteria for metabolic syndrome [49]. Similarly, a study by Thom et al. [50] indicated that the prevalence of MetS in Mexican Americans (31.9%) was significantly higher than that of non-Hispanic whites (23.8%).

The high prevalence of MetS among Hispanics demonstrates poor metabolic health across several measures and can also help to explain why Hispanics are 1.7 times more likely to have diabetes than non-Hispanic whites and 1.3 times more likely than non-Hispanic whites to die from diabetes [18]. In a population study of adults aged 45–84 years, 11.3% of Hispanics had a diagnosis of diabetes, compared to just 6.3% of non-Hispanic whites [51]. Another study found that Hispanic adults are 66% more likely to have diabetes relative to non-Hispanic white adults [52]. Schneiderman et al. [53] found differences in the prevalence rate of diabetes among a sample of 16,385 Hispanics from diverse backgrounds with Mexican (18.3%), Dominican and Puerto Rican (18.0%), Central American (17.0%) study participants having higher rates to Cuban (13.4%) and South American (10.2%).

Estimates suggest that 33.8% of the adult population of the United States meet criteria for obesity obese—defined as BMI $\geq$ 30 [54]. However, rates of obesity are significantly higher among Hispanics. Hispanics are 1.2 times more likely to be obese than non-Hispanic whites [19]. Data from the CDC [55] show that Hispanics (34.9%) have a higher prevalence of obesity than non-Hispanic white adults (30%). Rates among Hispanic women, specifically, are particularly elevated, with 35.4%

considered obese compared to 28.7% of the non-Hispanic white female population [55]. Differences between Hispanic men (34.3%) and non-Hispanic white men (31.2%) while not as large, but still highlight a significant disparity [55].

Regular physical activity has been shown to improve overall health and fitness, prevent the onset, or reduce the impact of many of the above-mentioned chronic diseases. Unfortunately, Hispanics are more likely to be sedentary, not engaged in leisure-time physical activity, or meet the recommended levels of physical activity compared to non-Hispanic white adults [44]. A study done by Neighbors et al. [45] found that while all Hispanic groups were less active than their non-Hispanic white counterparts, there was a lot of variability seen between Hispanic subgroups. Specifically, Cubans and Dominicans (19.8%) were the least active—defined as meeting recommended levels of physical activity set forth in *Healthy People 2020* [56]. In contrast, Mexicans were the most active (31.9% met physical activity recommendations; [45]).

MetS, diabetes, and obesity are prevalent among individuals with chronic mental illness [57], and the relationship between these cardiometabolic diseases and mental illness appears to be bidirectional. A meta-analysis conducted by Pan and colleagues (2014) found that those who were depressed were significantly more likely to develop MetS than those who were not depressed. Interestingly, individuals with MetS were significantly more at risk of developing depression than those without MetS. The same pattern was found for diabetes where depression appears to be both an outcome of, and a risk factor for the onset of diabetes. Rotella and Mannucci [58] conducted a meta-analysis encompassing 23 longitudinal studies and found that depression was associated with a 38% increased risk of diabetes. Other studies have found similar results. Depression comorbid with diabetes has been shown to significantly increase the risk of cardiovascular mortality and all-cause mortality [59, 60]. In addition, those with a chronic mental illness have a nearly four-fold greater risk of developing obesity than people without a previous psychiatric history [61–63]. The burden of mental illness comorbid with cardiometabolic diseases is disproportionate among Hispanics compared to non-Hispanic whites.

The prevalence of MetS in Hispanic adults with a chronic mental illness (74%) is significantly greater than their non-Hispanic white counterparts (41%; [64]). Furthermore, Cardenas et al. [65] conducted a study examining the links between depression and MetS in a sample of community-dwelling older Hispanics with type 2 diabetes. They found that depression was associated with a six-fold increase in risk for MetS. In addition, they investigated the relations between depression and specific MetS criteria, such as hypertension, dyslipidemia (as measured by HDL-C and plasma triglyceride levels), and obesity (as measured by BMI). They found that depression was significantly associated with elevated plasma triglyceride levels and low HDL-C. The authors speculate that depression may be interacting with diabetes to produce excess buildup of triglycerides in the bloodstream. Depression and diabetes may further interact to prevent HDL-C from transporting lipids to the liver to be broken down. Alternatively, it is possible that diabetic individuals with low HDL-C and high triglycerides may exacerbate the physical symptoms of depression (e.g., malaise), which, in turn, has an impact on health behaviors [65]. For example, Hispanics who have depression comorbid with diabetes may feel less motivated and have less energy to engage in adequate self-care, such as eating healthy and being physically active compared to non-depressed individuals. Regardless it appears that depression is associated with the physical activity as well as the biomarkers that play important roles for Hispanics in their level of risk for developing cardiometabolic diseases.

Active screening for depression among patients with diabetes should be a first-line approach for primary care providers. If depression is evident, referral of these patients to receive one or more evidence-based treatments (EBTs) such as anti-depressive medications or psychosocial treatments should be considered. Problem-solving therapy, behavioral activation, cognitive behavioral therapy, bibliotherapy, brief psychodynamic therapy, and reminiscence therapy have all been shown to be highly effective treatments for depression comorbid with a cardiometabolic disease [66]. It is important to note the sociocultural barriers that contribute to the disparities we see in mental health care in the Hispanic community [67–69]. Thus, it is imperative that any EBT be delivered within appropriate sociocultural contexts (e.g., delivering services in their primary language, address cultural health beliefs, and involve community health workers) and with affirming, person-centered language [70–72].

COVID-19

The novel coronavirus disease 2019 (COVID-19) pandemic has dramatically impacted the health and well-being of millions of individuals residing in the United States and around the world. A recent retrospective cohort found that one-third of individuals diagnosed with COVID-19 developed a neuropsychiatric disorder 6 months after its diagnosis. Among psychiatric conditions, anxiety was the most common sequelae (17.39%), followed by mood disorders (13.66%), substance use disorders (6.58%), and psychotic disorders (1.40%). In this same cohort, the rate of developing any neuropsychiatric disorder increased depending on the severity of COVID-19 (38.73% for hospitalized patients and 46.42% among patients receiving intensive care treatment; [73]). Hispanics, in particular, have been disproportionately impacted by the pandemic. At the time of writing, Hispanics account for one in three of all confirmed COVID-19 cases in the United States [16]. Furthermore, Hispanics have among the highest age-adjusted rates of COVID-19-associated hospitalizations at 117 per 100,000 and account for approximately one in five of all confirmed COVID-19-related deaths in the United States [16]. The impact that the pandemic has had on the mental health of the Hispanic population in the United States is yet to be elucidated; however, there is heightened concern that COVID-19 will exacerbate and amplify the mental health burden. Prevention and mitigation strategies (e.g., physical distancing, shelter-in-place, leaving home only for essential activities) help to attenuate adverse outcomes from COVID-19. These approaches may have unintended consequences on the physical and mental health of Hispanics given that Hispanics have identified the social support of family and close friends as an essential component to mental health and their well-being [71]. While the initial focus has been on understanding the virus itself, there are also significant public health challenges arising from the resulting prevention and mitigation measures as they are associated with increased cardiometabolic risk, depression, anxiety, isolation, loneliness and decreased quality of life.

Before the pandemic, insufficient physical activity was already described as a public health problem in the Hispanic community [44, 45] with social distancing measures exacerbating this disparity. Social isolation and loneliness have been correlated with levels of morbidity and mortality comparable to more established biopsychosocial risk factors like obesity, sedentary behavior, and hypertension [74, 75]. Given that familism and social cohesion are strong hallmarks of Hispanic culture [76, 77] and that Hispanics have high levels of stigmatizing attitudes towards mental illness, Hispanics may experience the harmful effects of comorbid social isolation and stigma more intensely than their non-Hispanic white counterparts. COVID-19 prevention and mitigation strategies such as physical distancing and shelter-in-place may worsen feelings of social isolation and loneliness. The awareness of vulnerability as well as the involuntary and inescapable self-isolation generate anxiety and other psychiatric symptoms of distress. While these measures are necessary for limiting COVID-19 cases, it seems evident that actions taken to curb spread of the virus will, in fact, exacerbate isolation-related vulnerabilities among Hispanics.

HIV

Although the incidence of HIV has begun to decline since the mid-2000s, rates of infection among certain demographic groups have continued to rise, particularly among Hispanic communities, sexual and gender minorities, and older adults. Although Hispanics have been significantly affected by HIV since the early years of the epidemic, these HIV-related disparities have strengthened over time. In the United States, Hispanics account for over 20% of the national HIV prevalence despite Hispanics making up less than 1/5 of the entire US population [78]. According to the most recent annual HIV/AIDS report from the CDC, there were 37,968 newly diagnosed Hispanics with HIV (HWH) in 2018—a 14% increase since 2010 [79]. These racial/ethnic disparities are further exacerbated when considering the experiences of Hispanic sexual minority men who make up nearly 25% of all new HIV infections [79] despite Hispanic men representing less than 9% of the entire US population. Additionally, less than 2/3 of HWH are engaged in routine care (58%) or virally suppressed (54%; [80])—two key indicators of HIV care management necessary for ending the HIV epidemic [81]. With these significant HIV disparities in mind, it is imperative for clinicians to be cognizant of the intersections between Hispanic individuals' race/ethnicity and other marginalized identities (e.g., age, sexual identity, immigration status) in developing and implementing culturally competent mental health interventions to best support HWH.

In addition to physical comorbidities, mental health disorders have a significant impact on the general health and quality of life of HWH. While data suggests that

HWH have comparable rates of psychiatric disorders compared to non-Hispanic people with HIV (PWH), self-reported psychiatric stress among HWH appears to be quite high [82]. Depression is the most prevalent comorbid mental health condition affecting PWH with data suggesting that anywhere between 33 and 47% of HWH are experiencing moderate to severe depression at any given time [82, 83]. Besides the anhedonia and negative mood that comes with depression, HWH with depression suffer from greater rates of treatment non-adherence [84, 85], more severe measures of disease progression [86], and even mortality [87]. Additionally, while there are multiple precipitants that lead to depression in HWH, research suggests that some of the more common factors such as the trauma associated with an HIV diagnosis [88] or more sociocultural factors such as social isolation, intersectional stigma, and discrimination [89, 90] may lead to these increased risks among HWH.

HWH who suffer from depression may be at increased risk for additional mental health challenges that negatively influence HWH's overall quality of life. PWH, including HWH, are five to ten times more likely to be living with a comorbid anxiety disorder compared to the general population [91] with evidence suggesting that PWH have higher rates of generalized anxiety disorder (5.6% compared to 2.5%), panic disorder (10.2% compared to 3%), and social anxiety disorder (9.1% compared to 6%) compared to HIV-negative peers [92]. For HWH in particular, high rates of anxiety disorder (35.4%) and post-traumatic stress disorder (24.2%) have been observed among HWH [93]; however, more research must be conducted that focuses predominantly on the prevalence of these two mental health diagnoses specifically among HWH.

Finally, in addition to mood and anxiety disorders, substance use and alcohol use disorders are common comorbidities for HWH. More generally, prevalence of substance use among PWH has been documented in the literature to range anywhere from 25% to 40% [94, 95] with a significant number of PWH self-endorsing polydrug use [95, 96]. Studies of HWH have observed rates of current hazardous alcohol use and substance use to be between 20 and 35% [93, 97] with lifetime prevalence rates of having ever engaged in binge drinking or substance use as high as 75% [97]. However, since other researchers have demonstrated less substance use or hazardous alcohol use among HWH compared to non-Hispanic PWH [98], more research is needed to gain more clarity among substance use among HWH. Current hazardous drinking or substance use has been associated with more acts of condomless sex [93], poor treatment adherence [99], and greater challenges in engaging with the HIV care continuum among HWH.

Although PWH in general have elevated rates of mental health challenges compared to their HIV-negative peers, HWH may be at even greater risk for experiencing psychological distress due to the additional stress associated with their various marginalized identities. Research indicates a strong link between HIV-related stigma, discrimination, and measures of psychological distress [100, 101]; therefore, the fact that HWH must navigate between various intersecting stigmas (e.g., racism, homophobia, HIV, immigration) may contribute to the high rates of mental health challenges among this population [102, 103]. Finally, various sociocultural factors associated with Hispanic culture (e.g., *machismo, fatalismo, familismo*, and maternal role as caretaker rather than care-seeker) may put HWH at an even greater risk for experiencing psychological distress [99, 104]. Since *familismo* and social cohesion are foundational pillars of Hispanic culture [76, 77], it is possible that HWH may experience more detrimental effects of comorbid stigma and social isolation from family and friends due to their status and isolation more intensely than non-Hispanic PWH; however, more research must be conducted to determine the veracity of such a hypothesis.

Tobacco Use

Tobacco use is a leading major public health issue. Cigarette smoking remains a leading cause of preventable disease and premature death in the United States and globally with other countries. On average, 435,000 people in the United States die prematurely from smoking-related diseases each year; overall, smoking causes 1 in 5 deaths. One out of five deaths attributed to cigarette smoking [105, 106]. Smoking is the leading cause of chronic conditions such as CVD (30% of heart disease death is caused by smoking) and lung cancer (the leading cause of cancer death in United States) and is associated with many other debilitating conditions such as cerebrovascular disease, chronic obstructive pulmonary disease, emphysema, peripheral vascular disease, and many types of cancers [107]. Individuals with mental illness are disproportionately affected by tobacco use compared to the general population. Almost one in two cigarettes in the United States are consumed by smokers with mental illness, and three out of four patients with a psychiatric illness are smokers. Unfortunately, tobacco use disorder is one of the most often ignored illness among physicians. While 69% of smokers want to quit, only 48% who saw a health professional the last year reported receiving advice to quit, and of those who tried to quit, only 31% received either counseling or medication. This is particularly true for psychiatrists, who have minimized, normalized, and rationalized its use among their patients. In a 2007 Survey of the American Association of Medical Colleagues, 47% of psychiatrists felt patients had more pressing problems to address and 22% reported that cessation would likely exacerbate comorbid psychiatric symptoms [107].

Rates of smoking are lower for Hispanics when compared to the general US population (9.8% vs 13.7 among adults and 3.8% vs 5.8% among high schoolers); however, it is important to note that there are broad variations among Hispanic subgroups, highlighting the need to disaggregate. For instance, according to data from the HCHS/SOL, Puerto Rican and Cuban men have higher rates of smoking prevalence, which exceeds the rates of non-Hispanic Whites (35% and 31.1%, respectively, vs 22.6% in 2010). On the other hand, Dominican and South American men have a tobacco use prevalence well below the national average (11% and 15.8%, respectively). Mexican and central American men have smoking rates comparable to non-Hispanic men (23.4% and 20.6%, respectively). In general, Hispanic women have lower rates of smoking than men at a rate approximately equal as non-Hispanic women [108].

Over the last 50 years, there has been a US-wide decline in smoking secondary to tobacco control efforts. For perspective, prevalence of smoking in 1970 was 50% compared to 18% today. Unfortunately, the impact of these smoking cessation campaigns has not been as effective among the Hispanic population. A 2017 survey to assess smoking prevention efforts in California, which has a strong anti-tobacco climate, showed that Whites had a 13.3% decrease in heavy smoking between the 1990s and the 2000s compared to a 5.1% decrease in heavy smoking among Mexican Americans and 2.0% among Central/South Americans, highlighting the need of further strategies to ensure Hispanics benefit from anti-tobacco measures equally [109]. Additionally, even though there has been a significant decrease in heavy smoking, this has been accompanied by a rising trend in light and intermittent smoking among Hispanics. Furthermore, Hispanic smokers have lower odds of receiving quitting advice from a health professional compared to their White counterparts [110].

Although this section of the chapter focuses primarily on cigarette smoking as way of administration, it is important to note there are many other forms of tobacco products and alternative nicotine delivery systems (such as smokeless or dissolvable tobacco). Special mention should be given to the use of cigars and cigarrillos with a prevalence of 12.8% among daily nicotine users according to the 2012–2013 National Adult Tobacco Survey (NATS). These products are notably more harmful than regular cigarettes, given the higher concentration of nicotine, toxins, and other carcinogens. This number might be much higher in certain Hispanics population, such as Cuban-American and other Caribbean-Americans, as it is notably part of their culture and history [108].

The reason for nicotine being so addictive is a result of psychological, biological, and even socioeconomic factors. First, nicotine is tremendously prone to secondary reinforcement; inhaled nicotine reaches the brain within 15 seconds, producing a sensation of pleasure, arousal, and a subjective feeling of relaxation, calmness and lessening of anxiety leading to a learned behavior of coping with stress via smoking. Second, the half-life of nicotine is only two hours, giving rise to a rapid desensitization of nicotine receptors, and a prompt precipitation of withdrawal characterized by irritability, frustration, difficulty concentrating, and restlessness. Withdrawal usually peaks in the first three days, and can last up to three weeks, with other worsening symptoms such as anxiety, depression, and insomnia. Third, the lack of an intoxication syndrome alongside the fact that is legal, permits smokers to use higher amounts of nicotine without the feel of it interfering with their daily functioning. Finally, the nicotine industry has specifically and heavily targeted vulnerable population with marketing strategies, such as the mentally ill, youth, and minorities (i.e., Hispanics; [107]).

Contrary to common belief that tobacco cessation will exacerbate psychiatric symptoms, there is overwhelming evidence demonstrating significant improvement in mental health and quality of life, including decreases in the severity of anxiety, depression, and stress alongside an overall improvement in mood among individuals with psychiatric illness who quit smoking. Moreover, cessation of use of nicotine products actually enhances that likelihood of achieving sobriety from other addictive substances, such as alcohol, opioids, and cocaine. Every individual should be screened for tobacco use in every psychiatric or medical visit, assessed for their motivation to quit, and assist in treatment, if indicated [107].

There are currently seven FDA-approved pharmacological interventions for smoking cessation and an even larger quantity of psychosocial interventions. A combination of both augments the chance of success by about 50%. Non-pharmacological interventions that have proven efficacious include motivational interviewing, cognitive behavioral training such as feedback (e.g., CO meter scores) skills training, relapse prevention, and stimulus control, mindfulness or relaxation-based interventions, and a large number of free and easily accessible community resources (e.g., hotlines, websites, twelve-step facilitation groups, peer specialists, etc.; [107]).

Pharmacological interventions include five nicotine replacement therapies which can be obtained over the counter (e.g., gums, lozenges, patch) or with a prescription (nicotine inhaler and nasal spray) and two medications, bupropion (a norepinephrine-dopamine receptor antagonist) and varenicline (an A2B4 partial ligand), all aimed to reduce cravings and withdrawal symptoms. All pharmacotherapies are equally effective in mono-therapy head-to-head trials except for varenicline, which has shown clinical and significant superiority.

For individuals looking to quit smoking, a reasonable strategy would be to start either bupropion or varenicline one to two weeks before a set quit date, followed by NRT upon quitting (e.g., daily patch with as needed gum/lozenges for cravings). It is generally recommended to continue maintenance therapy for six months and to gradually taper the medication afterwards over another six months. Additionally, as mentioned before, combining pharmacological interventions with psychosocial strategies enhances sobriety. There is much controversy regarding setting non-abstinence goals (reducing tobacco intake without reaching complete sobriety). Data suggests this may be beneficial, as this increases the likelihood of quitting at a later date. To our knowledge there are no randomized clinical trials that have been designed to assess pharmacotherapy strictly for Hispanics.

Health Behaviors

While psychotropic medications (e.g., antidepressants, benzodiazepines, antipsychotics) are the most widely used modality for treating mental illness, the clinical benefits of these medications must be balanced against their risks. Specifically, antipsychotics use is associated with significant weight gain, hyperglycemia, and lipid abnormalities while benzodiazepine use in older adults is associated with an increased risk of falls, hip fracture, cognitive impairment, all-cause mortality, overdose, and substance use disorder. Additionally, Hispanics are less likely to find antidepressant medication acceptable than non-Hispanic whites [111] since many Hispanics view the use of traditional mental health services as highly stigmatizing [71]. In addition, available mental health treatments may not match the preferences, values, and beliefs of Hispanic patients, which can lead to the decision to not even initiate mental health treatment in the first place [71]. In a study by Carpenter-Song et al. [112], Hispanics with mental illness resented diagnostic labels that carry the risk of social rejection. In contrast, those who conceptualized mental illness as a problem with nerves or as a sickness that one has, much like a physical ailment, experienced less stigma [112]. This helps explain why many Hispanics tend to express psychological distress as somatic complaints (e.g., heaviness in the chest, dizziness, drowning), and why they are skeptical of the utility of traditional forms of mental health treatment [112].

Previous studies have consistently found that among different Hispanic groups mental illness was perceived to have a reciprocal relationship with chronic physical illnesses [113, 114]. In these studies, Hispanics believed that the onset of a chronic physical illness, such as diabetes, was precipitated by an extreme emotional or stressful event. The inability to cope with these emotions presented a barrier to self-management and limited their ability to engage in self-management behaviors. Conversely, the diagnosis and management of a chronic physical illness, such as diabetes, had a significant impact on their emotional health. Thus, it is necessary to develop alternative approaches that use a culturally sensitive, nonstigmatizing approach to improve the physical and mental health of this population. The salience of the chronic physical illness in participants' conceptualizations of mental illness suggests that health promotion interventions—defined as behavioral interventions that use counseling strategies to equip participants with the necessary knowledge and skills to modify and sustain a healthy diet, increased physical activity, and/or healthy weight—are well-aligned with their perceived needs and may provide a tangible treatment to address these needs.

Psychosocial strategies such as health promotion interventions can play a critical role in treating Hispanics with comorbidities. Health promotion interventions are behaviorally activating, reduce vulnerability factors, and may be more desirable for reasons of safety and patient preference. Studies have consistently shown that increased physical activity effectively reduces symptoms of depression and anxiety in older adults. Benefits of health promotion also include improvements in systemic inflammation, physical fitness, and quality of life [115–118]. Studies of large population cohorts have consistently shown an association between physical inactivity and low-grade systemic inflammation and interventional studies have shown a reduction of inflammatory markers following physical activity [119]. Thus, health promotion interventions may be less stigmatizing and culturally relevant while enhancing general well-being and bringing mental health benefits to Hispanics faced with health-related challenges.

References

1. Ariyo AA, Haan M, Tangen CM, Rutledge JC, Cushman M, Dobs A, Furberg CD. Depressive symptoms and risks of coronary heart disease and mortality in elderly Americans. Cardiovascular health study collaborative research group. Circulation. 2000;102(15):1773–9.

2. Porensky EK, Dew MA, Karp JF, Skidmore E, Rollman B, Shear MK, Lenze EJ. The burden of late-life generalized anxiety disorder: effects on disability, health-related quality of life, and healthcare utilization. Am J Geriatr Psychiatr. 2009;17:473–82.
3. Taylor WD, McQuoid DR, Krishnan KR. Medical comorbidity in late-life depression. Int J Geriatr Psychiatry. 2004;19(10):935–43.
4. Moussavi S, Chatterji S, Verdes E, Tandon A, Patel V, Ustun B. Depression, chronic diseases, and decrements in health: results from the world health surveys. Lancet. 2007;370(9590):851–88.
5. Fiske A, Wetherell JL, Gatz M. Depression in older adults. Annu Rev Clin Psychol. 2009;5:363–89.
6. Gallegos-Carrillo K, García-Peña C, Mudgal J, Romero X, Durán-Arenas L, Salmerónet J. Role of depressive symptoms and comorbid chronic disease on health-related quality of life among community-dwelling older adults. J Psychosom Res. 2009;66(2):127–35.
7. Unützer J, Schoenbaum M, Katon WJ, Fan M-Y, Pincus HA, Hogan D, Taylor J. Healthcare costs associated with depression in medically ill fee-for-service Medicare participants. J Am Geriatr Soc. 2009;57(3):506–10.
8. Vogeli C, Shields AE, Lee TA, Gibson TB, Marder WM, Weiss KB, Blumenthal D. Multiple chronic conditions: prevalence, health consequences, and implications for quality, care management, and costs. J Gen Intern Med. 2007;22(Suppl 3):391–5.
9. Yang Y, George LK. Functional disability, disability transitions, and depressive symptoms in late life. J Aging Health. 2006;17:263–92.
10. Alegría M, Canino G, Shrout PE, Woo M, Duan N, Vila D, Torres M, Chen CN, Meng XL. Prevalence of mental illness in immigrant and non-immigrant U.S. Latino groups. Am J Psychiatry. 2008;165(3):359–69. https://doi.org/10.1176/appi.ajp.2007.07040704.
11. Breslau J, Aguilar-Gaxiola S, Kendler KS, Su M, Williams D, Kessler RC. Specifying race-ethnic differences in risk for psychiatric disorder in a USA national sample. Psychol Med. 2006;36(1):57–68. https://doi.org/10.1017/S0033291705006161.
12. Kessler RC, Ustun TB. The world mental health (WMH) survey initiative version of the World Health Organization (WHO) composite international diagnostic interview (CIDI). Int J Methods Psychiatr Res. 2004;13(2):93–121. https://doi.org/10.1002/mpr.168.
13. Jimenez DE, Alegría M, Chen C, Chan D, Laderman M. Prevalence of psychiatric illnesses in older ethnic minority adults. J Am Geriatr Soc. 2010;58(2):256–64.
14. Centers for Disease Control and Prevention. Health United States. 2019:2018. https://www.cdc.gov/nchs/data/hus/hus18.pdf.
15. Centers for Disease Control and Prevention. Summary health statistics: national health interview survey: 2018. 2020a. http://www.cdc.gov/nchs/nhis/shs/tables.htm.
16. Centers for Disease Control and Prevention. (2021b). Demographic trends of COVID-19 cases and deaths in the US reported to CDC. https://covid.cdc.gov/covid-data-tracker/?CDC_AA_refVal=https%3A%2F%2Fwww.cdc.gov%2Fcoronavirus%2F2019-ncov%2Fcases-updates%2Fcases-in-us.html#demographics
17. Daviglus ML, Talavera GA, Avilés-Santa ML, Allison M, Cai J, Criqui MH, et al. Prevalence of major cardiovascular risk factors and cardiovascular diseases AMONG HISPANIC/LATINO individuals of diverse backgrounds in the United States. JAMA. 2012;308(17):1775. https://doi.org/10.1001/jama.2012.14517.
18. Office of Minority Health. (2018). Diabetes and Hispanic Americans. https://www.minorityhealth.hhs.gov/omh/browse.aspx?lvl=4&lvlid=63
19. Office of Minority Health. (2020). Obesity and Hispanic Americans. https://www.minorityhealth.hhs.gov/omh/browse.aspx?lvl=4&lvlid=70#:~:text=Among%20Hispanic%20American%20women%2C%2078.8,obese%20than%20non%2DHispanic%20whites
20. González HM, Tarraf W, Whitfield KE, Vega WA. The epidemiology of major depression and ethnicity in the United States. J Psychiatr Res. 2010;44(15):1043–51. https://doi.org/10.1016/j.jpsychires.2010.03.017.
21. Bahrami H, Kronmal R, Bluemke DA, Olson J, Shea S, Liu K, Burke GL, Lima JAC. Differences in the incidence of congestive heart failure by ethnicity: the multi-ethnic

study of atherosclerosis. Arch Intern Med. 2008;168:2138–45. https://doi.org/10.1001/archinte.168.19.2138.
22. Rodriguez CJ, Allison M, Daviglus ML, Isasi CR, Kellr C, Leira, E.C, Sims, M. Status of cardiovascular disease and stroke in Hispanics/Latinos in the United States: a science advisory from the American Heart Association. Circulation. 2014;130:593–625. https://doi.org/10.1161/CIR.0000000000000071.
23. Vivo RP, Krim SR, Krim NR, Zhao X, Hernandez AF, Peterson ED, et al. Care and outcomes of Hispanic patients admitted with heart failure with preserved or reduced ejection fraction: findings from get with the guidelines-heart failure. Circ Heart Fail. 2012;2012(5):167–75. https://doi.org/10.1161/CIRCHEARTFAILURE.111.963546.
24. Shapiro PA, Critchfield AR. Cardiovascular disorders. In: Kaplan & Sadock's comprehensive textbook of psychiatry, vol. 2. 10th ed; 2017. p. 2186–200. Chapter, Wolters Kluwer.
25. Abramson J, Berger A, Krumholz HM, Vaccarino V. Depression and risk of heart failure among older persons with isolated systolic hypertension. Arch Intern Med. 2001;161(14):1725. https://doi.org/10.1001/archinte.161.14.1725.
26. Freedland KE, Carney RM, Rich MW. Effect of depression on prognosis in heart failure. Heart Fail Clin. 2011;7(1):11–21. https://doi.org/10.1016/j.hfc.2010.08.003.
27. Elfassy T, Al Hazzouri Z, Cai J, Baldoni PL, Llabre MM, Rundek T, et al. Incidence of hypertension among US Hispanics/Latinos: the Hispanic community health study/study of Latinos, 2008 to 2017. J Am Heart Assoc. 2020;9(12):e015031. https://doi.org/10.1161/jaha.119.015031.
28. Shaw PM, Chandra V, Escobar GA, Robbins N, Rowe V, Macsata R. Controversies and evidence for cardiovascular disease in the diverse Hispanic population. J Vasc Surg. 2018;67(3):960–9. https://doi.org/10.1016/j.jvs.2017.06.111.
29. Sorlie PD, Allison MA, Aviles-Santa ML, Cai J, Daviglus ML, Howard AG, et al. Prevalence of hypertension, awareness, treatment, and Control in the Hispanic community health study/study of Latinos. Am J Hypertens. 2014;27(6):793–800. https://doi.org/10.1093/ajh/hpu003.
30. Glassman AH, O'Connor CM, Califf RM, Swedberg K, Schwartz P, Bigger JT Jr, Krishnan KR, van Zyl LT, Swenson JR, Finkel MS, Landau C, Shapiro PA, Pepine CJ, Mardekian J, Harrison WM, Barton D, Mclvor M. Sertraline Antidepressant Heart Attack Randomized Trial (SADHEART) Group. Sertraline treatment of major depression in patients with acute MI or unstable angina. *JAMA*. 2002;288(6):701–9. https://doi.org/10.1001/jama.288.6.701. Erratum in: JAMA 2002 Oct 9;288 (14):1720
31. Kim J-M, Stewart R, Lee Y-S, Lee H-J, Kim MC, Kim J-W, et al. Effect of escitalopram vs placebo treatment for depression on long-term cardiac outcomes in patients with acute coronary syndrome. JAMA. 2018;320(4):350. https://doi.org/10.1001/jama.2018.9422.
32. Go AS, Mozaffarian D, Roger VL, Benjamin EJ, Berry JD, Borden, W.B.,…Turner, M.B. Heart disease and stroke statistics–2013 update: a report from the American Heart Association. Circulation. 2013;127:e6–e245.
33. Robinson RG, Jorge RE. Post-stroke depression: a review. Am J Psychiatr. 2016;173(3):221–31. https://doi.org/10.1176/appi.ajp.2015.15030363.
34. Moon JR, Glymour MM. Stroke incidence in older US Hispanics: is foreign birth protective? (S13.006). Neurology. 2012;78:1224–9. (Meeting abstracts 1). https://doi.org/10.1212/wnl.78.1_meetingabstracts.s13.006.
35. Sozener CB, Lisabeth LD, Shafie-Khorassani F, Kim S, Zahuranec DB, Brown DL, et al. Trends in stroke recurrence in Mexican Americans and non-Hispanic whites. Stroke. 2020;51(8):2428–34. https://doi.org/10.1161/strokeaha.120.029376.
36. Morgenstern LB, Smith MA, Lisabeth LD, Risser JM, Uchino K, Garcia N, Longwell PJ, McFarling DA, Akuwumi O, Al-Wabil A, Al-Senani F, Brown DL, Moyé LA. Excess stroke in Mexican Americans compared with non-Hispanic whites: the brain attack surveillance in Corpus Christi project. Am J Epidemiol. 2004 Aug 15;160(4):376–83. https://doi.org/10.1093/aje/kwh225.
37. Smith MA, Risser JM, Moyé LA, Garcia N, Akiwumi O, Uchino K, Morgenstern LB. Designing multi-ethnic stroke studies: the brain attack surveillance in Corpus Christi (BASIC) project. Ethn Dis. 2004. Autumn;14(4):520–6.

38. McCutcheon ST, Robinson RG. Neuropsychiatric aspects of cerebrovascular disorders. In: Kaplan & Sadock's comprehensive textbook of psychiatry, vol. 1. 10th ed; 2017. p. 474–89. Chapter, Wolters Kluwer.
39. Hackett ML, Anderson CS, House A, Xia J. Interventions for treating depression after stroke. Cochrane Database Syst Rev. 2008; https://doi.org/10.1002/14651858.cd003437.pub3.
40. Robinson RG, Jorge RE, Moser DJ, Acion L, Solodkin A, Small SL, et al. Escitalopram and problem-solving therapy for Prevention of Poststroke depression. JAMA. 2008;299(20):2391. https://doi.org/10.1001/jama.299.20.2391.
41. Jorge RE, Robinson RG, Arndt S, Starkstein S. Mortality and Poststroke depression: a placebo-controlled trial of antidepressants. Am J Psychiatr. 2003;160(10):1823–9. https://doi.org/10.1176/appi.ajp.160.10.1823.
42. Angel RJ, Angel JL. Minority group status and healthful aging: Social structure still matters. Am J Public Health. 2006;96:1152–9. https://doi.org/10.2105/AJPH.2006.085530.
43. Beard HA, Al Ghatrif M, Samper-Ternent R, Gerst K, Markides KS. Trends in diabetes prevalence and diabetesrelated complications in older Mexican Americans from 1993–1994 to 2004–2005. Diabetes Care. 2009;32:2212–7. https://doi.org/10.2337/dc09-0938.
44. Cromwell SL, Berg JA. Lifelong physical activity patterns of sedentary Mexican American women. Geriatr Nurs. 2006;27(4):209–13.
45. Neighbors CJ, Marquez DX, Marcus BH. Leisure-time physical activity disparities among Hispanic subgroups in the United States. Am J Public Health. 2008;98(8):1460–4.
46. World Health Organization. Definition, diagnosis, and classification of diabetes mellitus and its complications: report of a WHO Consultation. Part 1: diagnosis and classification of diabetes mellitus, 1999. Retrieved from http://whqlibdoc.who.int/hq/1999/WHO_NCD_NCS_99.2.pdf
47. Arenillas JF, Moro MA, Dávalos A. The metabolic syndrome and stroke: potential treatment approaches. Stroke. 2007;38:2196–203.
48. Aguilar M, Bhuket T, Torres S, Liu B, Wong RJ. Prevalence of the metabolic syndrome in the United States, 2003-2012. JAMA. 2015;313:1973–4.
49. Heiss G, Snyder ML, Teng Y, Schneiderman N, Llabre MM, Cowie C, et al. Prevalence of metabolic syndrome among Hispanics/Latinos of diverse background: the Hispanic community health study/study of Latinos. Diabetes Care. 2014;37(8):2391–9.
50. Thom T, Haase N, Rosamond W, Howard VJ, Rumsfeld J, Manolio T, et al. Heart disease and stroke statistics–2006 update: a report from the American heart association statistics committee and stroke statistics subcommittee. Circulation. 2006;113:e85–151.
51. Nettleton JA, Steffen LM, Ni H, Liu K, Jacobs DR. Dietary patterns and risk of incident type 2 diabetes in the multi-ethnic study of atherosclerosis (MESA). Diabetes Care. 2008;31:1777–82.
52. Roger VL, Go AS, Lloyd-Jones DM, Adams RJ, Berry JD, Brown, T.M.,… Wylie-Rosett, J. Heart disease and stroke statistics–2011 update: a report from the American Heart Association. Circulation. 2011;123:e18–e209.
53. Schneiderman N, Llabre M, Cowie CC, Barnhart J, Carnethon M, Gallo, L.C.,… Avilés-Santa, M.L. Prevalence of diabetes among Hispanics/Latinos from diverse backgrounds: the Hispanic community health study/study of Latinos (HCHS/SOL). Diabetes Care. 2014;37:2233–9.
54. Flegal KM, Carroll MD, Ogden CL, Curtin LR. Prevalence and trends in obesity among US adults, 1999-2008. JAMA. 2010;303(3):235–41. https://doi.org/10.1001/jama.2009.2014.
55. Centers for Disease Control and Prevention. E-cigarette use among middle and high school students–United States, 2020. Centers for Disease Control and Prevention. 2020, September 17. https://www.cdc.gov/mmwr/volumes/69/wr/mm6937e1.htm
56. United States Department of Health and Human Services. Healthy People 2020. Washington, DC, USA: United States Department of Health and Human Services; 2010.
57. Hellerstein DJ, Almeida G, Devlin MJ, Mendelsohn N, Helfand S, Dragatsi D, et al. Assessing obesity and other related health problems of mentally ill Hispanic patients in an urban outpatient setting. Psychiatry Q. 2007;78:171–81.

58. Rotella F, Mannucci E. Depression as a risk factor for diabetes: a meta-analysis of longitudinal studies. J Clin Psychiatry. 2013;74:31–7.
59. Park M, Katon WJ, Wolf FM. Depression and risk of mortality in individuals with diabetes: a meta-analysis and systematic review. Gen Hosp Psychiatry. 2013;35(3):217–25.
60. van Dooren FE, Nefs G, Schram MT, Verhey FR, Denollet J, Pouwer F. Depression and risk of mortality in people with diabetes mellitus: a systematic review and meta-analysis. PLoS One. 2013;8:e57058.
61. Consensus development conference on antipsychotic drugs and obesity and diabetes. Diabetes Care. 2004;27(2):596–601.
62. Henderson DC. Schizophrenia and comorbid metabolic disorders. J Clin Psychiatry. 2005;66(Suppl 6):11–20.
63. Saari KM, Lindeman SM, Viilo KM, Isohanni MK, Järvelin MR, Laurén LH, Savolainen MJ, Koponen HJ. J Clin Psychiatry. 2005;66(5):559–63.
64. Kato MM, Currier MB, Gomez CM, Hall L, Gonzalez-Blanco M. Prevalence of metabolic syndrome in Hispanic and non-Hispanic patients with schizophrenia. Primary Care Companion to the Journal of Clinical Psychiatry. 2004;6:74–7.
65. Cardenas V, Mausbach BT, Sommerfeld D, Jimenez DE, von Känel R, Ho JS, Garcia P, Aarons G. Depression is associated with increased risk for metabolic syndrome in Latinos with type 2 diabetes. Am J Geriatr Psychiatr. 2017;25(6):646–53.
66. Scogin FR, Moss K, Harris GM, Presnell AH. Treatment of depressive symptoms in diverse, rural, and vulnerable older adults. Int J Geriatr Psychiatry. 2014;29:310–6.
67. Harris KM, Edlund MJ, Larson S. Racial and ethnic differences in the mental health problems and use of mental health care. Med Care. 2005;43:775–84.
68. Jimenez DE, Bartels SJ, Cardenas V, Alegría M. Stigmatizing attitudes towards mental illness among racial/ethnic older adults in primary care. Int J Geriatr Psychiatry. 2013a;28(10):1061–8.
69. Jimenez DE, Cook B, Bartels SJ, Alegría M. Disparities in mental health service use of racial and ethnic minority elderly adults. J Am Geriatr Soc. 2013b;61:18–25.
70. Ingram M, Torres E, Redondo F, Bradford G, Wang C, O'Toole ML. The impact of promotoras on social support and glycemic control among members of a farmworker community on the US-Mexico border. Diabetes Educator. 2007;33(Suppl 6):172S–8S.
71. Jimenez DE, Bartels SJ, Cardenas V, Daliwal SS, Alegría M. Cultural beliefs and mental health treatment preferences of ethnically diverse older adult consumers in primary care. Am J Geriatr Psychiatr. 2012;20:533–42.
72. Jimenez DE, Syed S, Perdomo-Johnson D, Signorile JF. ¡HOLA, amigos! Toward preventing anxiety and depression in older Latinos. Am J Geriatr Psychiatr. 2018;26(2):250–6.
73. Taquet M, Geddes JR, Husain M, Luciano S, Harrison PJ. 6-month neurological and PSYCHIATRIC outcomes in 236 379 survivors of covid-19: a retrospective cohort study using electronic health records. Lancet Psychiatry. 2021;8(5):416–27. https://doi.org/10.1016/s2215-0366(21)00084-5.
74. Holt-Lunstad J, Smith TB, Layton JB. Social relationships and mortality risk: a meta-analytic review. PLoS Med. 2010;7:e1000316. https://doi.org/10.1371/journal.pmed.1000316.
75. Luanaigh CO, Lawlor BA. Loneliness and the health of older people. Int J Geriatr Psychiatry. 2008;23(12):1213–21. https://doi.org/10.1002/gps.2054.
76. Miyawaki CE. Association of social isolation and health across different racial and ethnic groups of older Americans. Ageing Soc. 2015;35(10):2201–28. https://doi.org/10.1017/S0144686X14000890.
77. Tomaka J, Thompson S, Palacios R. The relation of social isolation, loneliness, and social support to Disease outcomes among the elderly. J Aging Health. 2006;18(3):359–84. https://doi.org/10.1177/0898264305280993.
78. Centers for Disease Control and Prevention. HIV and older Americans. Centers for disease control and prevention. 2020b, Sep 14. https://www.cdc.gov/hiv/group/age/olderamericans/index.html.

79. Centers for Disease Control and Prevention. HIV and gay and bisexual Men. Centers for disease control and prevention. 2020c, Sept 16. https://www.cdc.gov/hiv/group/msm/index.html
80. Gant Z, Bradley H, Hu X, Skarbinski J, Hall HI, Lansky A, Centers for Disease Control and Prevention (CDC). Hispanics or Latinos living with diagnosed HIV: progress along the continuum of HIV care–United States, 2010. MMWR Morb Mortal Wkly Rep. 2014;63(40):886–90.
81. 90–90-1: treatment for all | UNAIDS. 2013. Retrieved January 27, 2021, from https://www.unaids.org/en/resources/909090
82. Shapiro MF, Morton SC, McCaffrey DF, Senterfitt JW, Fleishman JA, Perlman JF, Athey LA, Keesey JW, Goldman DP, Berry SH, Bozzette SA, Additional Authors From the HCSUS Consortium. Variations in the care of HIV-infected adults in the United States: results from the HIV cost and services utilization study. JAMA. 1999;281(24):2305. https://doi.org/10.1001/jama.281.24.2305.
83. Cook JA, Cohen MH, Burke J, Grey D, Anastos K, Kirstein L, Palacio H, Richardson J, Wilson T, Young M. Effects of depressive symptoms and mental health quality of life on use of highly active antiretroviral therapy among HIV-seropositive women. JAIDS Journal of Acquired Immune Deficiency Syndromes. 2002;30(4):401–9. https://doi.org/10.1097/01.QAI.0000018280.16783.FE.
84. Levison JH, Levinson JK, Alegría M. A critical review and commentary on the challenges in engaging HIV-infected Latinos in the continuum of HIV care. AIDS Behav. 2018;22(8):2500–12. https://doi.org/10.1007/s10461-018-2187-1.
85. Simoni JM, Wiebe JS, Sauceda JA, Huh D, Sanchez G, Longoria V, Andres Bedoya C, Safren SA. A preliminary RCT of CBT-AD for adherence and depression among HIV-positive Latinos on the U.S.-Mexico border: the Nuevo Día study. AIDS Behav. 2013;17(8):2816–29. https://doi.org/10.1007/s10461-013-0538-5.
86. Leserman J. Role of depression, stress, and trauma in HIV Disease progression. Psychosom Med. 2008;70(5):539–45. https://doi.org/10.1097/PSY.0b013e3181777a5f.
87. Cook JA, Grey D, Burke J, Cohen MH, Gurtman AC, Richardson JL, Wilson TE, Young MA, Hessol NA. Depressive symptoms and AIDS-related mortality among a multisite cohort of HIV-positive women. Am J Public Health. 2004;94(7):1133–40. https://doi.org/10.2105/AJPH.94.7.1133.
88. Sauceda JA, Wiebe JS, Chan K, Kutner BA, Simoni JM. Acculturation, family cohesion, and mental health among Latinos living with HIV on the U.S.-Mexico border. Cultur Divers Ethnic Minor Psychol. 2018;24(3):453–8. https://doi.org/10.1037/cdp0000166.
89. Algarin AB, Zhou Z, Cook CL, Cook RL, Ibañez GE. Age, sex, race, ethnicity, sexual orientation: intersectionality of marginalized-group identities and enacted HIV-related stigma among people living with HIV in Florida. AIDS Behav. 2019;23(11):2992–3001. https://doi.org/10.1007/s10461-019-02629-y.
90. Earnshaw VA, Bogart LM, Dovidio JF, Williams DR. Stigma and racial/ethnic HIV disparities: moving toward resilience. Stigma and Health. 2015;1(S):60–74. https://doi.org/10.1037/2376-6972.1.S.60.
91. Pence BW, Miller WC, Gaynes BN, Eron JJJ. Psychiatric illness and Virologic response in patients initiating highly active antiretroviral therapy. JAIDS Journal of Acquired Immune Deficiency Syndromes. 2007;44(2):159–66. https://doi.org/10.1097/QAI.0b013e31802c2f51.
92. Brandt CP, Paulus DJ, Garza M, Lemaire C, Norton PJ, Zvolensky MJ. A novel integrated cognitive-behavioral therapy for anxiety and medication adherence among persons living with HIV/AIDS. Cogn Behav Pract. 2018;25(1):105–18. https://doi.org/10.1016/j.cbpra.2017.01.007.
93. Kutner BA, Nelson KM, Simoni JM, Sauceda JA, Wiebe JS. Factors associated with sexual risk of HIV transmission among HIV-positive Latino men who have sex with men on the U.S.-México border. AIDS Behav. 2017;21(3):923–34. https://doi.org/10.1007/s10461-016-1449-z.
94. Korthuis PT, Josephs JS, Fleishman JA, Hellinger J, Himelhoch S, Chander G, Morse EB, Gebo KA. Substance abuse treatment in human immunodeficiency virus: the role

of patient–provider discussions. J Subst Abus Treat. 2008;35(3):294–303. https://doi.org/10.1016/j.jsat.2007.11.005.
95. Sohler NL, Wong MD, Cunningham WE, Cabral H, Drainoni M-L, Cunningham CO. Type and pattern of illicit drug use and access to health care services for HIV-infected people. AIDS Patient Care STDs. 2007;21(s1):S-68. https://doi.org/10.1089/apc.2007.9985.
96. Proeschold-Bell RJ, Heine A, Pence BW, McAdam K, Quinlivan EB. A cross-site, comparative effectiveness study of an integrated HIV and substance use treatment program. AIDS Patient Care STDs. 2010;24(10):651–8. https://doi.org/10.1089/apc.2010.0073.
97. Rajabiun S, Rumptz MH, Felizzola J, Frye A, Relf M, Yu G, Cunningham WE. The impact of acculturation on Latinos' perceived barriers to HIV primary care. Ethn Dis. 2008;18(4):403–8. https://www.ncbi.nlm.nih.gov/pmc/articles/PMC4844059/
98. Luna-Gierke RE, Shouse RL, Luo Q, Frazier E, Chen G, Beer L. Differences in characteristics and clinical outcomes among Hispanic/Latino men and women receiving HIV medical care–United States, 2013–2014. Morb Mortal Wkly Rep. 2018;67(40):1109–14. https://doi.org/10.15585/mmwr.mm6740a2.
99. Campo RE, Alvarez D, Santos G, Latorre J. Antiretroviral treatment considerations in Latino patients. AIDS Patient Care STDs. 2005;19(6):366–74. https://doi.org/10.1089/apc.2005.19.366.
100. Herek GM, Saha S, Burack J. Stigma and psychological distress in people with HIV/AIDS. Basic Appl Soc Psychol. 2013;35(1):41–54. https://doi.org/10.1080/01973533.2012.746606.
101. Stutterheim SE, Pryor JB, Bos AE, Hoogendijk R, Muris P, Schaalma HP. HIV-related stigma and psychological distress: the harmful effects of specific stigma manifestations in various social settings. AIDS. 2009;23(17):2353–7. https://doi.org/10.1097/QAD.0b013e3283320dce.
102. Crenshaw K. Demarginalizing the intersection of race and sex: a black feminist critique of antidiscrimination doctrine, feminist theory and antiracist politics. Univ Chic Leg Forum. 1989;1989:139–68. https://heinonline.org/HOL/P?h=hein.journals/uchclf1989&i=143.
103. English D, Rendina HJ, Parsons JT. The effects of intersecting stigma: a longitudinal examination of minority stress, mental health, and substance use among black, Latino, and multiracial gay and bisexual men. Psychol Violence. 2018;8(6):669–79. https://doi.org/10.1037/vio0000218.
104. Galvan FH, Bogart LM, Wagner GJ, Klein DJ, Chen Y-T. Conceptualisations of masculinity and self-reported medication adherence among HIV-positive Latino men in Los Angeles, California, USA. Cult Health Sex. 2014;16(6):697–709. https://doi.org/10.1080/13691058.2014.902102.
105. Benowitz NL. Nicotine addiction. N Engl J Med. 2010;362(24):2295–303.
106. Doll R, Peto R, Boreham J, Sutherland I. Mortality in relation to smoking: 50 years' observations on male British doctors. Br Med J. 2004;328:1519.
107. Ziedonis DM, Tonelli ME, Das S. Tobacco-related disorders. In: Kaplan & Sadock's Comprehensive textbook of psychiatry, vol. 1. 10th ed; 2017. p. 1342–52. Essay, Wolters Kluwer.
108. Blanco L, Garcia R, Pérez-Stable EJ, White MM, Messer K, Pierce JP, Trinidad DR. National Trends in smoking behaviors among Mexican, Puerto Rican, and Cuban men and women in the United States. Am J Public Health. 2014;104(5):896–903. https://doi.org/10.2105/ajph.2013.301844.
109. Felicitas-Perkins JQ, Sakuma K-LK, Blanco L, Fagan P, Pérez-Stable EJ, Bostean G, et al. Smoking among Hispanic/Latino nationality groups and whites, comparisons between California and the United States. Nicotine Tob Res. 2017;20(9):1085–94. https://doi.org/10.1093/ntr/ntx191.
110. Kaplan RC, Bangdiwala SI, Barnhart JM, Castañeda SF, Gellman MD, Lee DJ, Pérez-Stable EJ, Talavera GA, Youngblood ME, Giachello AL. Smoking among U.S. Hispanic/Latino adults: the Hispanic community health study/study of Latinos. Am J Prev Med. 2014;46(5):496–506. https://doi.org/10.1016/j.amepre.2014.01.014.

111. Givens JL, Houston TK, Van Voorhees BW, Ford DE, Cooper LA. Ethnicity and preferences for depression treatment. Gen Hosp Psychiatry. 2007;29(3):182–91.
112. Carpenter-Song E, Chu E, Drake RE, Ritsema M, Smith B, Alverson H. Ethno-cultural variations in the experience and meaning of mental illness and treatment: implications for access and utilization. Transcult Psychiatry. 2010;47(2):224–51.
113. Cabassa LJ, Hansen MC, Palinkas LA, Ell K. Azúcar y nervios: explanatory models and treatment experiences of Hispanics with diabetes and depression. Soc Sci Med. 2008;66(12):2413–24.
114. Pincay IEM, Guarnaccia PJ. "It's like going through an earthquake": anthropological perspectives on depression among Latino immigrants. J Immigr Minor Health. 2007;9:17–28.
115. Barbour KA, Blumenthal JA. Exercise training and depression in older adults. Neurobiology and Aging. 2005;26(Suppl 1):119–23.
116. Blumenthal JA, Babyak MA, Doraiswamy PM, Watkins L, Hoffman BM, Barbour KA, et al. Exercise and pharmacotherapy in the treatment of major depressive disorder. Psychosom Med. 2007;69(7):587–96.
117. Martinsen EW. Physical activity in the prevention and treatment of anxiety and depression. Nord J Psychiatry. 2008;62(Suppl 47):25–9.
118. Ströhle A. Physical activity, exercise, depression and anxiety disorders. J Neural Transm. 2009;116(6):777–84.
119. Tafet GE, Nemeroff CB. The links between stress and depression: psychoneuroendocrinological, genetic, and environmental interactions. Journal of Neuropsychiatry and Clinical Neuroscience. 2016;28(2):77–88.

Mental Health in Hispanic/Latina/Latinx Women

14

Wilma I. Castilla-Puentes, Ruby Castilla-Puentes, and Sandra Castilla-Puentes

Introduction

Latinx people have become the largest minority group in the United States, making up about 17% of the population [1]. The heterogeneity of people who are Hispanic and Latinos, who are comprised of a variety of ethnic backgrounds such as Native American, African, and Spanish, makes identifying mental health needs a challenging but rewarding experience. The National Alliance on Mental Health reported that one in five Latinx people suffer from mental illness, making culturally competent outreach and treatment an urgent issue [2].

As part of the fastest growing and largest ethnic minority group, projections indicate that 128 million Latinos will reside in the United States by 2050 [1, 3]. Research on racial and ethnic disparities in healthcare access and utilization constantly identifies Latinos as one of the most disadvantaged ethnic groups. Using measures such as usual source of care, health insurance coverage, and the quality of care received, barriers for Latinas are readily identified [4]. However, an inclusive understanding of health and healthcare disparities must consider gender differences, given that health and illnesses are experienced differently by men and women. If demographic trends

W. I. Castilla-Puentes (✉)
Hospital de Sogamoso, Sogamoso, Boyaca, Colombia

WARMI Mental Health, Collaborative Mental Health Network, Boyaca, Colombia

R. Castilla-Puentes
Janssen Research & Development LLC, Titusville, NJ, USA

Center for Clinical and Translational Science and Training, University of Cincinnati, Cincinnati, OH, USA

WARMI Mental Health, Collaborative Mental Health Network, Cincinnati, OH, USA

S. Castilla-Puentes
WARMI Mental Health, Collaborative Mental Health Network, Boyaca, Colombia

Hospital Regional de Duitama, Duitama, Boyaca, Colombia

R. Castilla-Puentes, T. Falcone (eds.), *Mental Health for Hispanic Communities*,
https://doi.org/10.1007/978-3-031-13195-0_14

Table 14.1 Percentage of serious psychological distress among persons 18 years of age and over, by gender

Gender	Hispanic	Non-Hispanic White	Hispanic/Non-Hispanic White Ratio
Men	3.5	2.8	1.3
Women	5.5	4.8	1.1
Total	4.6	3.8	1.1

Source: CDC 2021. Summary Health Statistics: National Health Interview Survey: 2018. Table A-8. https://www.cdc.gov/nchs/nhis/shs/tables.htm

continue, it is suggested that Latinas will represent 25% of the total female population in the United States and make up 52% of the growing Latino population [5]. As a fast-growing demographic, it is important to address the challenges and barriers that may affect the quality of health and healthcare among Latina women. The terms Latino and Hispanic are used interchangeably in research to describe this fast-growing demographic [6]. Those who identify as Latino, Latina, or Latinx may consider themselves of Latin American ancestry (Central America, South America, or the Caribbean). Since the Spanish language is typically gendered, the term Latinx is used to eliminate a binary choice (male vs. female) that is limiting and excluding to trans individuals and others who identity as fluid or non-binary. Those who identify as Hispanic may be referring to ancestors from Spain or other Spanish-speaking [7].

Hispanic/Latina/Latinx Mental Health Data Related to Women in United States

Most of the current mental health studies combine minority groups—including Latina women—into a pool of research that generalizes their experience [8]. With the combination of ethnic groups, it is difficult to focus on one single population. A report at the US National Library of Medicine suggests that lifetime psychiatric disorder prevalence estimates were 28.1% for Hispanic men and 30.2% for Hispanic women. Increased rates of psychiatric disorders were observed among US-born, English-language-proficient, and third-generation Latinos [9]. The prevalence of serious mental illness is almost 70% greater in women than in men. Table 14.1 provides the percentage of serious psychological distress by gender.

Below is an overview of the most frequent psychiatric conditions affecting women in the United States.

Depression

Nadeem et al. [8] conducted a qualitative study with 15,383 participants who were low-income Latina and Black women, screened from Women Entering Care (WCE). They examined whether stigma was associated with seeking care or mental health treatment. Depression is a stigmatized illness, which makes it difficult for women to accept their condition and seek help. During the interviews, many were concerned about being characterized as "crazy," which decreased their disposition to employ in mental health services. Nadeem et al. [8] revealed that Latina women with a mental health illness were less likely to want treatment because of the stigma. This research lacked a certain percent of different mental illnesses that were presented in the study, especially a few women who were diagnosed with depression. The forthcoming study identifies Mexican women's presenting problem, diagnosis, and treatment that have improved their well-being. Zayas [10] interviewed 148 Hispanic and African American women who assessed to have elevated depressive symptoms. The study found that Hispanic women had received less social support and support networks than African American women [11]. This finding is startling since family is an important element in the Hispanic culture; however, the study describes that Hispanic women had less social support due to having family in another country. In the qualitative study, 50–75% of women who had depressive symptoms went undiagnosed and untreated [11]. This study, research suggested physicians to be familiar with depressive symptoms, however, did not encourage Latina women with depressive symptoms to seek preventive mental healthcare [11].

In addition, the amount of depression in Latino women compared to Latin men is much higher—46% compared to 19.6%. However, the American Psychiatric Association's Office of Minority and National Affairs suggests that among Hispanics with a mental disorder, less than 1/11 contacts a mental health specialist, and less than 1/5 contacts a general healthcare provider.

Latinx women are twice as likely to develop depression as compared to Latinx men, white populations, or African American populations [12]. Research also indicates that employed Latinx women are more stressed than unemployed ones. Findings show that this could be due to the added responsibilities that come with being a mother and working multiple jobs. Another factor regarding employment includes the frustration and depression that arises from Latinx women being overqualified for the jobs they work, due to racial and gender discrimination. Latinas are typically paid just 57 cents for every dollar paid to white, non-Hispanic men [13]. This gap in pay—which typically amounts to a loss of $2409 every month, $28,911 every year, and $1,156,440 over a 40-year career—means that Latinas must work over 21 months to make as much as white, non-Hispanic white men were paid in just 12. In 2019, 15.2% of adult Latinas lived in poverty. That rate worsened to 16.8% in 2020 [14].

Because of Latinas' higher rates of unemployment and economic insecurity, when they return to the workforce, many Latinas will be willing to accept the first job offer they receive because they cannot afford to be out of work any longer; employers, in turn, may pay lower wages to employees who have been unemployed or out of the workforce for long stretches of time. And because Latinas face such a steep gender wage gap, they were less able to afford education or professional training during the pandemic that would have allowed them to advance or move into another field. All of this could translate to larger race and gender wage gaps for Latinas moving forward.

Mental health statistics show that Hispanic women in the United States experience depression at about twice the rate of Hispanic males and are at a higher risk for depression than Caucasian and African American women [15]. This is, partially, due to multiple social determinants of health that impact Hispanic women and their families. Social determinants of health (i.e., education, income, health status, and acculturation) among Hispanic women may play a crucial role in the development or exacerbation of depression [16].

In the Latino culture, a woman's role is to informally maintain a strong foundation that keeps the family united. Meanwhile, the man role is to formally manage the family's basic needs [17].

Suicide

Suicidal behavior among Hispanic youth has been reported to be higher in comparison to non-Hispanic Black and White youth (Centers for Disease Control and Prevention, 2003, 2006; Substance Abuse and Mental Health Services Administration, 2003). Although data do not distinguish Latino youth by country of origin or heritage, Latino youth of both sexes have shown consistently higher rates of suicidal ideation, plans, and behavior than their non-Hispanic counterparts, except for the category of youth designated as "Other" (which includes Native American youth). Despite the higher than average reports of suicidal behavior among Hispanic youth, their actual rates of suicides are lower than those of non-Hispanic White and Native American adolescents, but higher than those of non-Hispanic Black youth [11].

Latin American culture is, traditionally, very family oriented but it retains firm gender divisions that celebrate achievements made by males while at the same time relegating women to roles of homemaking. Because these divisions run deep while at the same time families play such a strong part in the lives of Hispanics, young Latina girls feel guilt by not following their patriarchal demands. Or, if they decide to pursue their own ambitions regardless, they will create inner turmoil by following their dreams and going against family wishes [18].

In 2019, suicide was the second leading cause of death for Hispanics, ages 15 to 34 [19]. Suicide attempts for Hispanic girls, grades 9–12, were 30% higher than for non-Hispanic white girls in the same age group, in 2019 [19]. The high rates of suicidal behavior by teenage Hispanic females reported in large-scale surveys can be understood as a cultural phenomenon, a product of specific elements of the history, tradition, ideology, or social norms of a particular society, and that treatment interventions must take family and cultural factors into consideration [20]. A summary with the % of suicidal ideation among students who attempted suicide in grades 9–12 is noted in Table 14.2.

Table 14.2 Suicidal ideation (%) among students who attempted suicide in grades 9–12, 2018

	Hispanic	Non-Hispanic White	Hispanic/Non-Hispanic White Ratio
Men	5.5	6.4	0.9
Women	11.9	9.4	1.3
Total	8.9	7.9	1.1

Source: CDC 2021. High School Youth Risk Behavior Survey Data. [Accessed 04/21/2021]. https://nccd.cdc.gov/youthonline

Post-partum Depression (PPD)

Post-partum depression (PPD), a major depressive episode that occurs following delivery or within the first four weeks after delivery [21], is the leading medical complication among new mothers, and affects an estimated 12–19% of the general population of new mothers [22]. The prevalence of PPD in Latinas in the United States has been estimated at three to four times greater (30–43%) than the general population of new mothers [23]. A growing body of research suggests that psychosocial stressors are important risk factors for PPD [24]. There is evidence in the literature that prolonged exposure to stress is associated with hyperactivity of stress response systems, such as the hypothalamic-adrenal-pituitary (HPA) axis, which can become dysregulated over time, making it difficult for the individual to adapt to later stressors [25]. Dysregulation of the HPA axis has been implicated in the development of PPD [26]. However, the biological mechanisms by which psychosocial stressors are associated with an increased risk of PPD are poorly understood, particularly among immigrant and US-born Latinas who experience high rates of complex and long-lasting psychosocial stressors, because they are not well-represented in biomedical research and the multifaceted stressors they encounter throughout their life have not been examined [27].

While pregnant women everywhere are susceptible to PPD and other mental illnesses, the problem is greatest in developing countries, where WHO estimates that 20% of mothers experience post-partum depression. Latinas are part of the fastest growing minority group in the United States. In 2010, Latinas accounted for 24% of all births and immigrant Latinas accounted for 56% of births that year [28]. Despite the rapid growth of this population and high fertility rates among US- and foreign-born Latinas, there is limited information about the mental health of these women following delivery.

In a group of 116 Latinx immigrants, the prevalence of significant symptoms of PPD was 54.2% for the entire sample of 116 women. Nearly 66% of women who screened positive for symptoms of PPD scored above the listed cutoff score for suicidal thoughts [29]. PPD is exacerbated by poverty, migration, stress, and exposure to violence, according to research compiled by WHO. The organization emphasizes the need to integrate maternal mental health with general health guidelines, along with educating women about children's health and reproductive health. Latinas experience cultural and contextual contributors for PPD that should be assessed simultaneously along with biological factors, such as the HPA axis, that are affected by high levels of stress. This study has implications for clinical practice. Healthcare providers should routinely assess the presence, proximity, and degree of the contributors discussed here when evaluating immigrant and US-born post-partum Latinas for mood disorders. Similarly, health professionals should consider the potential effects of how a dysregulated HPA stress response might impact a woman's mood. In sum, it is vital that healthcare providers consider the broad range of cultural, contextual, and biological contributors to PPD noted here. [27].

Anxiety

The NIMH defines General Anxiety Disorder (GAD) as experiencing "excessive anxiety or worry" for most days over a period of 6 months. Other anxiety disorders include panic disorder, obsessive-compulsive disorder, social anxiety disorder (or social phobia), separation anxiety disorder, and phobia-related disorders (such as fear of flying, fear of heights, or fear of specific objects). While 19% of all adults in the US report having experienced anxiety disorder in the past year, the percentage is much higher for women than for men (23.4% vs. 14.3%).

Symptoms of anxiety disorder include the following: chronic irritability or nervousness, feelings of impending doom or disaster, racing heartbeat, hyperventilating, sweating, or trembling, weakness or tiredness, inability to concentrate, sleeplessness, stomach aches or other digestive problems.

Women are twice as likely as men to be impacted by Generalized Anxiety Disorder (GAD). The Anxiety and Depression Association of America (ADAA) reports that 6.8 million people in the United States are affected by GAD, although only 43% of them are being treated for the disorder. Women are also twice as likely as men to be diagnosed with panic disorder (PD), which affects six million US adults, and with specific phobias, which impact 19 million adults in the United States.

Post-traumatic Stress Disorders

Post-traumatic stress disorder (PTSD) is a serious and common mental illness with lifetime prevalence rates in the United States ranging from 3.4% to 17.7% [30], depending on sampling methods. Women are more likely to experience PTSD, and they wait much longer than men after symptoms arise to seek diagnosis and treatment. The Office of Women's Health at the U.S. Department of Health and Human Services reports that women wait an average of 4 years after the onset of PTSD symptoms before asking for help. Men, on the other hand, seek assistance an average of 1 year after PTSD symptoms arise. Sexual violence is the primary source of PTSD worldwide. Recovery Across Mental Health states that women have a higher rate of developing PTSD after a traumatic event: 20.4% for women, compared to 8.1% for men. According to the ADAA, 65% of male rape victims and 45.9% of women who are victims of rape will develop PTSD as a result [30].

Substance Abuse

The size of the Latina/Latino population in the United States and projections for its growth require that substance abuse researchers and treatment providers pay greater attention to the prevalence and treatment of drug and alcohol abuse among this group [31].

In general, Latinas are more likely than Latinos to abstain from using alcohol and illicit drugs and Latinas are less likely to drink heavily and become dependent on alcohol [9].

Substance abuse prevalence rates for Latinos/Latinas generally mirror those of the general US population; however, several indicators of assimilation to US culture as well as sociodemographic variables predict substance use and abuse among this group. Latinos/Latinas have poorer outcomes in substance abuse treatment programs. Yet there is little empirical evidence that explains the problems these individuals experience in treatment, and there are few studies on the use and effectiveness of mutual help groups among this population. New developments in the conceptualization and measurement of acculturation will lead to a greater understanding of the role of culture in the prevalence and treatment of substance-related problems [9].

Research from [9] provides some potential reasons for a higher rate of substance abuse in Hispanic females including a family history of substance abuse, mood and anxiety disorders, acculturation issues, unmarried, unemployment, exposure to traumatic events, and poverty [9].

Risk Factors for Mental Health Issues in Women

Studies suggest that the longer a Hispanic woman lives in the United States, the higher the risk for depressive symptoms, as there is an increased sense of loss in cultural values, norms, and family cohesion [16]. A key study into the social determinants of depression among Hispanic women suggests that when a Hispanic woman does not live with her partner, has an educational level of below high school, and has a fair or poor health, she has a significantly higher risk of developing depression [32]. Exposure to violence makes a woman three to four times more likely to be affected by depression. Researchers at WHO state that women who were exposed to sexual abuse as children, or to a violent partner as an adult, are diagnosed with depression at a much higher rate. The research also found that the severity and duration of the initial sexual or violent exposure impacts the severity of the resulting mental illness.

Women disproportionately experience the following risk factors for common mental disorders than men: 1) Women earn less than men; Hispanic/Latina/Latinx mental health data related to women in US risk factors for women who are full time workers earn about one-fourth less than male counterparts in a given year [33]; 2) the poverty rate for women aged 18 to 64 is 14.2% compared with 10.5% for men. For women aged 65 and older the poverty rate is 10.3%, while the poverty rate for men aged 65 and older is 7.0%; 3) victims of violence: About 1 in 3 women have experienced sexual violence, physical violence, and/or stalking by an intimate partner in their lifetime [34]; 4) an estimated 65% of caregivers are women. Female caregivers may spend as much as 50% Hispanic/Latina/Latinx mental health data related to women in US risk factors for more time providing care than male caregivers [7].

Insights for Clinicians

The presence of psychosocial stressors in many women's lives and its connection to depression and anxiety disorders warrants further research and evaluation for adequate social and financial supports. Follow up with proposed solutions when needed. Consider working with women in a collaborative care model, working together with a patient's primary care provider, and ancillary providers to deliver more holistic, comprehensive care. Be aware of biases that may come into play in the care of female patients, especially when there are symptoms with unique presentations.

Culture affected aspects of all the above themes, with birthplace sometimes modifying these effects. Regarding the physician–patient relationship, for instance, many women placed a high value on a caring social interaction whether they were born inside or outside the United States. Even so, birthplace (i.e., US born vs foreign born) seemed to influence some women's attitudes and preferences. Studies suggested that women who grew up in the United States differed from those born outside the United States in their emphases on aspects of the patient–physician relationship and communication. Some women born in the United States gave the impression that they regarded their doctor's role more as that of a paid professional, even though they still wished for a relationship characterized by warmth and compassion [35]. Women born outside the United States, however, tended to trust the doctor's medical training and automatically respect him or her as the authority in charge of their and their families' health. What they most desired from the relationship was the physician's ability to empathize with and understand them.

Conclusion and Future Needs

Hispanic/Latina/Latinx women, like other women have different needs for mental health services throughout their life. To respond to these needs it is necessary that during the training period of health professional's mental health as it relates to gender be included as a relevant aspect of their academic formation. With respect to the evaluation of health problems in women, several variables have been considered and emphasis has been placed on several psychosocial conditions that are common to most of Hispanic/Latinx. It is necessary to educate women, since the lack of education and poverty associated with cultural misconceptions maintains their poor life conditions, which in tum become psychosocial risk factors for mental disorders. Job conditions for Hispanic/Latinx women must improve, with access to social protection. Health services for women must integrate the mental health component in reproductive health services.

References

1. United States Census Bureau. Facts for features: hispanic heritage month 2017. Release Number CB17-FF.17 2017. https://www.census.gov/newsroom/facts-for-features/2017/hispanic-heritage.html
2. Moreno E. Latinos and mental health: current issues and a reason for hope. American Foundation for suicide Prevention. 2017. Retrieved from: https://afsp.org/latinos-mental-health-current-issues-reason-hope/

3. Valenzuela JM, McDowell T, Cencula L, Hoyt L, Mitchell MJ. Hazlo bien! A participatory needs assessment and recommendations for health promotion in growing Latino communities. Am J Health Promot. 2013 May-Jun;27(5):339–46.
4. Berdahl TA, Torres Stone RA. Examining Latino differences in mental healthcare use: the roles of acculturation and attitudes towards healthcare. Community Ment Health J. 2009 Oct;45(5):393–403.
5. Ramos B, Jurkowski J, Gonzalez B, Lawrence C. Latina women: health and healthcare disparities. Soc Work Public Health. 2010;25(3/4):258–71.
6. Jacobson L. The Hispanic dynamic: how will the rapidly growing Latino population affect politics and policy? State Legis. 2015;41(6):16.
7. SAMHSA. National survey on drug use and health 2016. 2015.
8. Nadeem E, Lange JM, Edge D, Fongwa M, Belin T, Miranda J. Does stigma keep poor young immigrant and US-born black and Latina women from seeking mental health care? Psychiatr Serv. 2007;58(12):1547–54.
9. Alvarez J, Jason LA, Olson BD, Ferrari JR, Davis MI. Substance abuse prevalence and treatment among Latinos and Latinas. J Ethn Subst Abus. 2007;6(2):115–41.
10. Zayas LH, Cunningham M, McKee MD, Jankowski KRB. Depression and negative life events among pregnant african-american and hispanic women. Women's Health Issues. 2002;12(1):16–22, ISSN 1049–3867. https://doi.org/10.1016/S1049-3867(01)00138-4.
11. Zayas LH, Pilat AM. Suicidal behavior in Latinas: explanatory cultural factors and implications for intervention. Suicide Life Threat Behav. 2008;38(3):334–42.
12. American Psychological Association, Latina Women Firsts. Ethnicity and health in America Series: Latina women's firsts in psychology. 2015. https://www.apa.org/pi/oema/resources/ethnicity-health/latino-heritage/women-firsts. Accessed 10/21/2021.
13. National Women's Law Center (NWLC) calculations using U.S. Census Bureau, Current Population Survey. Annual social and economic supplement. 2021. https://www.census.gov/data/tables/time-series/demo/income-poverty/cps-pinc/pinc-05.html
14. NWLC calculations using U.S. Census Bureau. Age and Sex of All People, Family Members and Unrelated Individuals. Table POV-01 in 2020 Current Population Survey and 2021 Current Population Survey. 2021, Sep 14. https://www.census.gov/data/tables/time-series/demo/income-poverty/cps-pov/pov-01.html
15. Shattell MM, Smith KM, Quinlan-Colwell A, Villalba JA. Factors contributing to depression in latinas of mexican origin residing in the United States: implications for nurses. J Am Psychiatr Nurses Assoc. 2008 Jun;14(3):193–204.
16. Lorenzo-Blanco EI, Cortina LM. Latino/a depression and smoking: an analysis through the lenses of culture, gender, and ethnicity. Am J Community Psychol. 2013 Jun;51(3–4):332–46.
17. Nuñez A, González P, Talavera GA, Sanchez-Johnsen L, Roesch SC, Davis SM, Gallo LC. Machismo, Marianismo, and negative cognitive-emotional factors: findings from the Hispanic community health study/study of Latinos sociocultural ancillary study. J Latina/O Psychol. 2016;4(4):202–17.
18. Landale NS, Oropesa RS, Bradatan C. Hispanic families in the United States: family structure and process in an era of family change. In: National Research Council (US) panel on Hispanics in the United States. In: Tienda M, Mitchell F, editors. Hispanics and the future of America, vol. 5. Washington (DC): National Academies Press (US); 2006. Available from: https://www.ncbi.nlm.nih.gov/books/NBK19902/.
19. CDC. National center for injury prevention and control. Web-based injury statistics query and reporting system (WISQARS). 2019. https://www.cdc.gov/injury/wisqars/index.html. Accessed 10/21/2021.
20. Nolle AP, Gulbas L, Kuhlberg JA, Zayas LH. Sacrifice for the sake of the family: expressions of familism by Latina teens in the context of suicide. Am J Orthopsychiatry. 2012;82(3):319–27.
21. American Psychiatric Association. Diagnostic and statistical manual of mental disorders (DSM-5®). American Psychiatric Pub. 2013.
22. Gavin NI, Gaynes BN, Lohr KN, Meltzer-Brody S, Gartlehner G, Swinson T. Perinatal depression: a systematic review of prevalence and incidence. Obstet Gynecol. 2005;106(5 Pt 1):1071–83.
23. Lucero NB, Beckstrand RL, Callister LC, Sanchez Birkhead AC. Prevalence of postpartum depression among Hispanic immigrant women. J Am Acad Nurse Pract. 2012;24(12):726–34.

24. Liu CH, Tronick E. Rates and predictors of postpartum depression by race and ethnicity: results from the 2004 to 2007 new York City PRAMS survey (pregnancy risk assessment monitoring system). Matern Child Health J. 2013;17(9):1599–610.
25. Juruena MF. Early-life stress and HPA axis trigger recurrent adulthood depression. Epilepsy Behav. 2014;38:148–59.
26. Jolley SN, Elmore S, Barnard KE, Carr DB. Dysregulation of the hypothalamic-pituitary-adrenal axis in postpartum depression. Biol Res Nurs. 2007;8(3):210–22.
27. Lara-Cinisomo S, Wisner KL, editors. Perinatal depression among Spanish- speaking and Latin American women: a global perspective on detection and treatment. New York, NY: Springer; 2013.
28. Livingston G, Cohn D. U.S. birth rate falls to a record low; decline is greatest among immigrants social and demographic trends. Pew Research Center. 2012;
29. Callister LC, Beckstrand R, Corbett C. Postpartum depression and help seeking behaviors in immigrant Hispanic women. J Obstet Gynecol Neonatal Nurs. 2011;40:S108.
30. Sibrava NJ, Bjornsson AS, Pérez Benítez ACI, Moitra E, Weisberg RB, Keller MB. Posttraumatic stress disorder in African American and Latinx adults: clinical course and the role of racial and ethnic discrimination. Am Psychol. 2019;74(1):101–16.
31. Delgado M. Cross-cutting themes and recommendations. In: Delgado M, editor. Latinos and alcohol use/abuse revisited. New York: Haworth; 2005. p. 205–12.
32. De Oliveira G, Cianelli R, Gattamorta K, Kowalski N, Peragallo N. Social determinants of depression among Hispanic women. J Am Psychiatr Nurses Assoc. 2017;23(1):28–36.
33. SAMHSA. Past year mental disorders among adults in the United States: results from the 2008–2012 mental health surveillance study. CBHSQ DATA Review. 2014. samhsa.gov/data/sites/default/files/NSDUH-DR-N2MentalDis-2014-1/Web/NSDUH-DR-N2MentalDis-2014.htm
34. American Psychiatric Association. Best practice highlights female patients. 2017. psychiatry.org/psychiatrists/cultural-competency/treating-diversepatient-populations/working-with-women-patients. Accessed 10/21/2021
35. Julliard K, Vivar J, Delgado C, Cruz E, Kabak J, Sabers H. What Latina patients don't tell their doctors: a qualitative study. Ann Fam Med. 2008;6(6):543–9.

Index

A

Acculturation, 96, 114
- clinical scene, 91–93
- components of, 89
- definition of, 87
- in Hispanic communities, 90, 91
- management of, 93, 95
  - educational or formative, 93, 94
  - political management, 94
  - public health, 94, 96
  - social, legal and administrative rules, 94
- process and experience of, 88–90
- risk of suicidal ideation, 112
- SUD, 128, 129

Acculturation difficulty, 71
Acculturative stress hypothesis, 112
Active suicidal ideation (SI), among Hispanic and Latino communities
- prevalence, 106
- risk factors, 107

Acute exacerbations, 40
Affordable care act (ACA), 65, 178
Agoraphobia, 79, 80
AIDS, 38
Al Andalus, 11
Alprazolam, 82
Alzheimer's disease, 139
America
- contribution of African medicine in, 14
- medical and mental health care in, 13, 14

Amphetamines, 39
Anabolic-androgenic steroids, 39
Anhedonia, 34
Antidepressant, 57, 82, 187
Antiepileptics, 37
Antipsychotics, 40
Anxiety, 23, 42, 84, 155–157
- agoraphobia, 79, 80
- attack of nerves, 70
- diagnostic and testing issues, 68
- Diagnostic and Statistical Manual of Mental Disorders (DSM-5), 69
- fright, 70
- GAD, 80–82
- general considerations, 67
- measures, 68
- due to medical condition, 83
- nerves, 70
- panic disorder, 78, 79
- selective mutism, 74
  - cultural issues, 74, 75
  - diagnosis, 75
  - epidemiology, 75
  - treatment, 75
- separation anxiety, 72
  - cultural issues, 72
  - diagnosis, 73
  - epidemiology, 73
  - treatment, 73
- social environment, problems related to
  - acculturation difficulty, 71
  - social exclusion or rejection, 71
  - target of (perceived) adverse discrimination or persecution, 71, 72
- social phobia, 76, 77
- specific phobia, 76
- specified, 83
- substance/medication-induced anxiety disorder, 83
- unspecified, 83

Attention deficit hyperactivity disorder (ADHD), 160, 161

B

Baroflex mechanism, 141
Barriers to care, 144

R. Castilla-Puentes, T. Falcone (eds.), *Mental Health for Hispanic Communities*,
https://doi.org/10.1007/978-3-031-13195-0

Beck Anxiety Inventory, 81
Benzodiazepines, 74, 76, 77, 79, 82
Berry's orthogonal, two-factor model of acculturation, 89
Bipolar affective disorder (BPAD), 55, 56, 60
Bipolar disorder, 2, 24
Blood pressure variability (BPV), 142
Body dysmorphic disorder, 84
Brief psychotic disorder, 40
Bupropion, 195
Buspirone, 77

C
Cannabis, 39
Cardiometabolic diseases, 188–191
Cardiovascular disease (CVD), 143, 184–186
Center for Epidemiological Studies Depression Scale, CES-D10, 54
Cerebrovascular disease, 186–188
Children's health insurance program (CHIP), 65
Chlorpromazine, 40
Cigarette smoking, 194
Clonazepam, 77, 82
Cognitive-behavioral therapy (CBT), 40, 73, 77
Cognitive behavioral therapy (CBT) with cognitive restructuring, 76, 82
Cognitive impairment, 138
Cognitive rehabilitation, 40
Cognitive remediation, 40
Community-Based Adult Services (CBAS), 146, 147
Comorbid psychiatric illness, 107
Coronary artery disease (CAD), 184
Coronavirus disease 2019 (COVID-19) pandemic, 40, 148, 190
Corticosteroids, 39
Coyotes, 88
Creutzfeldt-Jakob disease, 38
Cultural competence, 175
Cultural Formulation Interview (CFI), 170
Cultural identity, 88
Cultural paradox, 132
Cultural sensitive services for Latinos, 169
- advocacy for development and maintenance, 177, 178
- benefits of implementation and protection, 175, 176
- challenges, 176, 177
- with cultural approach, 170–172
- development of Hispanic mental health services
  - education, 173, 174
  - research, 172, 173
- inclusion, 174, 175
- in United States, 170

Culturally And Linguistically Diverse (CALD), 177

D
Deferred Action for childhood arrivals (DACA), 65, 156
Delusions, 34–36
Depression, 23, 42, 92, 108, 190
- barriers in treatment, 56, 57
- cultural influence, manifestation inHispanics, 57, 58
- Hispanic/Latina/Latinx mental health data related to women in US, 206–208
- Latinx youth, 154, 155
- prevalence of
  - differences by ethnic group, 53–55
  - with other ethnicities living in USA, 50–53

Depressive symptomatology, 143, 144
Diagnostic and Statistical Manual of Mental Disorders, Revised Third Edition (DSM-III-R), 50
Diagnostic and Statistical Manual of Mental Health Disorders (DSM-5), 63, 69
Dichos, 176
Discrimination, 114, 118
Doxepin (Sinequan), 82
Drug-induced psychosis, 39
Dyslipidemia, 143
Dysthymic disorder, 50

E
Eating disorders, 21
Education level, 140
Electroconvulsive therapy (ECT), 40
Endocrine diseases, 38
Epilepsy, 37
Epsin 4 gene, 43
Escitalopram (Lexapro), 82, 187
Escitalopram for Depression in Acute Coronary Syndrome (EsDEPACS) study, 186
Estrogens, 39
Ethnic identity, 174
Ethnicity, 52, 138
Ethnoracial studies, 139
Evidence-based treatments (EBTs), 190
Excoriation disorder, 84

F
Faith Based Organizations (FBO), 3
Familism, 115, 131
Family/courtesy stigma, 23
Fluoxetine (Prozac), 74, 79
Fluphenazine, 40
Fluvoxamine (Luvox), 74, 77

G
Gabapentin, 77
Gender differences, 109–111
Generalized anxiety disorder (GAD), 80–82, 92, 183, 210
Genetic studies, 59, 60
Genome-wide analysis (GWA), 137
Gray matter atrophy, 139
Greek-Roman heritage, 10, 11
Group-oriented approach, 93

H
Hallucinations, 34–36
Hamilton Anxiety rating scale (Ham-A), 81
Health behaviors, 195, 196
Health literacy, 141, 144
Health promotion interventions, 196
Heart rate variability (HRV), 141
Hemorrhagic strokes, 37
Herpes simplex virus, 38
Heterogeneity, 205
Hispanic and Latino Groups
  Bipolar Affective Disorder, prevalence of, 55, 56
  composition, 49
  cultural influence of depression manifestation in, 57, 58
  difficulties studying minorities, 49, 50
  disinformation and stigma, 56, 57
  genetic studies, 59, 60
  prevalence of depression
    differences by ethnic group, 53–55
    with other ethnicities living in USA, 50–53
  psychopharmacogenetic studies, 58, 59
  in United States, 64
    anxiety (*see* Anxiety)
    cultural matters, 66, 67
    demographics, 64
    education, 64
    family values, 66
    health care, 65
    immigration status, 65
    income, 65, 66
    language, 66
    media, 66
    religion, 66
Hispanic communities
  acculturation, 90, 91
  anxiety (*see* Anxiety)
  mental health for, 1–5
  psychotic disorders in (*see* Psychotic disorders in Hispanics)
Hispanic Community Health Study/Study of Latinos (HCHS/SOL), 59
Hispanic paradox, 96, 138
Hispanic world, 9
  Greek-Roman heritage, 10, 11
Hispanic/Latina/Latinx mental health data related to women in US, 206, 212
  depression, 206–208
  insights for clinicians, 212
  PPD, 209, 210
  PTSD, 210
  risk factors for, 211
  substance abuse, 210, 211
  suicidal behavior, 208
  suicide, 208
HIV, 191–193
Hoarding disorder, 84
Hospitals creation, for mentally ill, 11–13
Hydroxyzine, 82
Hypertension, 186

I
Iberia, 9, 10
  Islamic medicine in, 11
Imipramine (Tofranil), 82
Immigrant paradox, 51, 183
Immigration, 114
Infectious illnesses, 37, 38
Inhalants, 39
Intersectional stigma, 18
Ischemic strokes, 37
Islamic medicine in Iberia, 11

K
Kluver-Bucy syndrome, 38

L
Latino Behavioral Health Services (LBHS), 171
Latino paradox, 106, 114, 118, 173
Lifetime substance use disorder (LT-SUD), 3

M

Machismo, 132
Major depression disorder (MDD), 59, 60
Major depressive episodes, 36
Major neurocognitive disorder (MaND), 145
Malignancies, 38
Marianismo, 132
Marijuana, 43, 128
Medicaid, 65, 157
Medical comorbidities, 184
- cardiometabolic diseases, 188–191
- cerebrovascular disease, 186–188
- CVD, 184–186
- health behaviors, 195, 196
- HIV, 191–193
- Tobacco use, 193–195

Mental health care providers, 26
Mental health for Hispanic communities, 1–5
Mental health in Hispanic youth, 153
- ADHD, 160, 161
- anxiety, 155–157
- depression in Latinx youth, 154, 155
- suicide, 158, 159

Mental health in Latinx youth, 161, 162
Mental health practitioners, 26, 27
Mental health stigma
- comparisons of, 19
- definition of, 17
- intersectional stigma, 18
- in Latinx community
  - culturally matched providers, increases, 27
  - factors, increases, 21–24
  - impacts on treatment, 20, 21
  - inclusion of family in treatment, 28
  - increasing Mental Health Knowledge, 27, 28
  - interventions to reduce, 24, 25
  - levels in, 18, 19
- perceived (public) stigma, 17
- self-stigma, 18
- types of, 17, 18

Mental illness stigma, *see* Mental health stigma
Metabolic diseases, 38
Metabolic syndrome (MetS), 188, 189
Migration status, 139, 140
Migratory phenomena, 88
Mood disorder, 36
Multiple housing programs, 65
Multiple sclerosis (MS), 37

N

National Epidemiologic Survey on Alcohol and Related Conditions (NESARC), 2
Neurologic diseases, 37
Neuroscience, crisis in 19th century and, 14, 15
Nicotine, 194
Nicotine replacement therapies, 195
Non-psychiatric comorbidities, 141–143
Non-suicidal self-injury (NSSI), 106
- among Hispanic and Latino communities, 111
  - prevalence, 110
  - risk factors, comorbidities, and gender differences, 110, 111
- definition of, 110

O

Obsessive compulsive disorder (OCD), 84
Opioid misuse, 128
Osteosarcoma, 5

P

Panic Attack Scale, 82
Panic attacks, 78, 79
Panic disorder (PD), 78, 79, 210
Panic-focused cognitive behavioral therapy (CBT), 79
Paraneoplastic syndromes, 38
Paroxetine (Paxil), 77, 79, 82
Passive suicidal ideation (SI)
- among Hispanic and Latino communities, 106
- definition of, 106

Perceived (public) stigma, 17
Perceived stigma, 18, 20–22
Pharmacotherapies, 195
Postpartum depression (PPD), 209, 210
Post-traumatic stress disorder (PTSD), 43, 84, 210
Pre-Columbian America, medical approach in, 13
Pregabalin, 77
Preguntas con Cartas method, 19
Problem solving therapy (PST), 187
Propranolol (Inderal), 77
Psychiatric comorbidity, 187
Psychiatric medication, 5
Psychoeducation, 40
Psychopharmacogenetic studies, 58, 59

Psychopharmacologic therapy, 40
Psychosis, 37
Psychotherapy, 4, 95
Psychotic disorders in Hispanics, 33
  among Hispanic Americans, 42–45
  brief psychotic disorder, 40
  drug-induced psychosis, 39
  endocrine and metabolic diseases, 38
  infectious illnesses, 37, 38
  malignancies, 38
  mood disorder, 36
  neurologic diseases, 37
  psychosis, 37
  schizoaffective disorder, 36
  schizophrenia, 34, 35
    criteria of DSM-5, 35
    differential diagnosis of, 35
    hereditability and family history of, 34
    neuropsychological assessment, 35
    pathophysiology of, 34
    symptoms, 34
  Schizophreniform disorder, 36
  treatment of, 40–42
Psychotropic medications, 22
Public stigma, 17, 21

R
Race, 51, 52, 138
Racial discrimination, 22
Racial/ethnic differences, 183
Racism, 22
Reasons For Living Inventory (RFLI), 114
Recommendations
  caring for the caregivers, 147
  Community-Based Adult Services (CBAS), 146, 147
  COVID-19 Pandemic, 148
  literacy campaigns, 144
  optimize medication adherence, 145
  participation in clinical trials, 145
  physical exercise and dietary adjustments, 146
  training of clinicians, 145
Rehabilitative therapy, 40
Religion and spirituality (R/S), 1–5
Religion, definition of, 2

S
Schizoaffective disorder, 36
Schizophrenia, 34, 35, 43
  criteria of DSM-5, 35
  differential diagnosis of, 35
  hereditability and family history of, 34
  neuropsychological assessment, 35
  pathophysiology of, 34
  prognosis, 42
  symptoms, 34
Schizophreniform disorder, 36
Secret feelings, 25, 26
Selective mutism, 74
  cultural issues, 74, 75
  diagnosis, 75
  epidemiology, 75
  treatment, 75
Selective Serotonin Reuptake Inhibitors (SSRIs), 73, 76
Self-rating scales, 82
Self-stigma, 18, 20
Separation anxiety, 72
  cultural issues, 72
  diagnosis, 73
  epidemiology, 73
  treatment, 73
Sertraline (Zoloft), 74, 77, 79
Simpatía, 131
Single nucleotide polymorphisms (SNPs), 43, 59
Smoking cessation, 195
Social anxiety, *see* Social phobia
Social exclusion/rejection, 71
Social phobia, 76, 77
Social rehabilitation, 77
Socio-demographic factors, 89
Socio-economic factors, 89
Socioeconomic status (SES), 141
Specific phobia, 76
Spiritualism, 132
Spirituality, 2
Stigma related to eating disorders, 20
Stroke, 37, 186
Structural stigma, 18
Subjective cognitive decline (SCD), 145
Substance abuse, 210, 211
Substance use disorders, 35, 43, 92, 127
  acculturation, role of, 128, 129
  cultural aspects for treating, 130–132
  definition of, 127
  differences in Latin American place of origin, 129
  gender and, 129
  in Latino population, 127
  treatment, 130
Substance/medication-induced anxiety disorder, 83

Suicidal behavior, 111, 118
evidence-based interventions, 116, 117
Hispanic/Latina/Latinx mental health data related to women in US, 208
protective factors among Hispanic and Latino communities, 114, 115
familism, 115
religion, 115, 116
in US-born vs. foreign-born Hispanics, 113, 114
Suicidal ideation, 208
Suicide, 105, 158, 159
acculturation, 112
attempts among Hispanic and Latino communities, 110
gender differences, 109
prevalence, 108, 109
risk factors and comorbidities, 109
Hispanic/Latina/Latinx mental health data related to women in US, 208
plans among Hispanic and Latino communities, 107
prevalence, 107
risk factors, gender, and comorbidities, 107, 108
Suicide-related behaviors, 106
Supplemental nutrition assistance program (SNAP), 65
Systemic lupus erythematosus (SLE), 38

T
Telephone Interview for Cognitive Status (TICS), 145
Temporal lobe atrophy, 139
Temporary Protected Status (TPS), 64
Tobacco use, 193–195
Trauma, 153
Trazodone (Desyrel), 82
Trichotillomania, 84
Tricyclics, 74

V
Venlafaxine (Effexor), 77
Vitamin B1 (thiamine) deficiency, 38
Vitamin B12 deficiencies, 38

W
Wernicke-Korsakoff syndrome, 38
White matter hyperintensity (WMH), 139

Batch number: 10732090

Printed by Printforce, the Netherlands